PERGAMON INTERNATIONAL LIBRARY
of Science, Technology, Engineering and Social Studies
*The 1000-volume original paperback library in aid of education,
industrial training and the enjoyment of leisure*
Publisher: Robert Maxwell, M.C.

Load-Bearing Fibre Composites

THE PERGAMON TEXTBOOK
INSPECTION COPY SERVICE

An inspection copy of any book published in the Pergamon International Library will gladly be sent to academic staff without obligation for their consideration for course adoption or recommendation. Copies may be retained for a period of 50 days from receipt and returned if not suitable. When a particular title is adopted or recommended for adoption for class use and the recommendation results in a sale of 12 or more copies, the inspection copy may be retained with our compliments. The Publishers will be pleased to receive suggestions for revised editions and new titles to be published in this important International Library.

INTERNATIONAL SERIES ON THE STRENGTH
AND FRACTURE OF MATERIALS AND STRUCTURES

General Editor: D. M. R. Taplin, D.Sc., D.Phil., F.I.M.

Other Titles in the Series

EASTERLING
Mechanisms of Deformation and Fracture

HAASEN AND GEROLD
Strength of Metals and Alloys (ICSMA 5) – 3 Volumes

MILLER AND SMITH
Mechanical Behaviour of Materials (ICM 3) – 3 Volumes

SMITH
Fracture Mechanics – Current Status, Future Prospects

TAPLIN
Advances in Research on the Strength and Fracture of
Materials – 6 Volumes

Related Pergamon Journals

Engineering Fracture Mechanics

Fatigue of Engineering Materials and Structures

NOTICE TO READERS

Dear Reader

If your library is not already a standing order customer or
subscriber to this series, may we recommend that you place a
standing or subscription order to receive immediately upon
publication all new issues and volumes published in this
valuable series. Should you find that these volumes no longer
serve your needs your order can be cancelled at any time
without notice.
The Editors and the Publisher will be glad to receive suggestions or outlines of suitable titles, reviews or symposia for
consideration for rapid publication in this series.

ROBERT MAXWELL
Publisher at Pergamon Press

Frontispiece. A recent application of the first glass-polymer: the 107 m long radome on top of the Toronto 550 m CN Tower (Courtesy of CN Tower Ltd.)

Load-Bearing Fibre Composites

BY

MICHAEL R. PIGGOTT

A.R.C.S., Ph.D, P.Eng., F.Inst.P.

Centre for the Study of Materials, University of Toronto

PERGAMON PRESS

OXFORD · NEW YORK · TORONTO · SYDNEY · PARIS · FRANKFURT

U.K.	Pergamon Press Ltd., Headington Hill Hall, Oxford OX3 OBW, England
U.S.A.	Pergamon Press Inc., Maxwell House, Fairview Park, Elmsford, New York 10523, U.S.A.
CANADA	Pergamon of Canada, Suite 104, 150 Consumers Road, Willowdale, Ontario M2J 1P9, Canada
AUSTRALIA	Pergamon Press (Aust.) Pty. Ltd., P.O. Box 544, Potts Point, N.S.W. 2011, Australia
FRANCE	Pergamon Press SARL, 24 rue des Ecoles, 75240 Paris, Cedex 05, France
FEDERAL REPUBLIC OF GERMANY	Pergamon Press GmbH, 6242 Kronberg-Taunus, Hammerweg 6, Federal Republic of Germany

First edition 1980

British Library Cataloguing in Publication Data

Piggott, M R
Load bearing fibre composites. – (Pergamon international library: international series on the strength and fracture of materials and structures).
1. Fibrous composites
620.1 1   TA418.9.C6   79–40951

ISBN 0–08–024230–8 (Hardcover)
ISBN 0–08–024231–6 (Flexicover)

Printed in Great Britain by A Wheaton & Co Ltd; Exeter

Contents

Preface

THIS book is based on a course of lectures given to upper-year and graduate students in Materials Science and Engineering. It is designed to present a unified view of the whole field of fibre (and platelet) composites, rather than going into any aspect of the subject in great detail. The reader who wants to go more deeply into any aspect is referred, at the end of each chapter, to more specialized texts and reviews, or to key papers. The reader is also recommended to look at the literature acknowledged in the subtitles to the figures.

The field has developed rapidly over the last 20 years, and no materials scientist should now be without a working knowledge of it. Metallurgists should be aware of the competition from reinforced polymers and ceramics, and be able to appreciate their strengths and weaknesses. Designers need to be able to make a rational choice of which material to use in any situation. This book is aimed at these audiences, but, in addition, indicates areas where our knowledge is not as complete as it should be. Thus I hope also to inspire those inclined to do research.

A mathematical development of the subject cannot be avoided if composites are to be properly understood and correctly used. The interactions between fibres and matrix are quite complex, and are the subject of multitudes of erudite papers. Here, however, complicated mathematical expressions are eschewed, and wherever simple ones can be used with adequate precision these are presented in their place.

All equations are developed from basic principles, and the reader only needs acquaintance with algebra, calculus, and geometry at first-year University level.

The author is grateful for assistance from many sources. Pictures and information have been generously provided, and the author has been kindly received in many laboratories where important work is going on, and benefited greatly from what he has seen and discussed there.

The writer is especially grateful for assistance while on study leave at the University of Bath. The University provided material assistance, and Professor Bryan Harris unstintingly provided original pictures, comments, encouragement, and assistance.

1

Introduction

THIS book describes the basic ideas in the relatively new field of fibre reinforced composites, and provides recent data on their load-bearing capabilities. In addition, selected data is given for more traditional materials, so that the reader can see where composite materials fit into the hierarchy of materials.

The discussion starts at a basic level, and such important properties as strength, creep resistance, and fatigue resistance are described. The connection between chemical bond strength and tensile strength is discussed, and the role of imperfections is shown to be of overwhelming importance.

The reasons for the good mechanical properties of fibres are set forth, and an account is given of the production of strong and stiff fibres. This is followed by a simplified description of the mechanics of the processes by which fibres can contribute strength, stiffness, and toughness to weak matrices. Finally a brief sketch of the properties of composites is given, classified according to whether the matrix is polymeric, metallic, or ceramic (including cementitious). Emphasis is given, throughout the book, to load-bearing properties and this is followed by a discussion of some of the diverse end uses of these materials.

The term composite has come to mean a material made by dispersing particles, of one or more materials in another material, which forms a substantially continuous network around them. The properties of the composite may bear little relation to those of the components, even though the components retain their integrity within the composite.

The components can be randomly arranged, or organized in some sort of pattern. Generally the arrangement will have a large effect on the properties. Further, they can have roughly spherical shapes, e.g. stones in concrete, or can have some very distinctive shape such as the iron carbide laminae found in some steels, or long thin fibres, such as the cellulose fibres in wood. The particle shape also has a very profound effect on the properties of the composite. Since this book is concerned with fibre-reinforced composites, it will largely restrict itself to the discussion of two-component composites, where one of the components is a long thin fibre or whisker, and where the composite has to have good load-carrying capacity. Also reinforcement by thin platelets will be described.

Fibre-reinforced materials have been used by man for a very long time. The first to be used were naturally occurring composites, such as wood, but man also found out, long ago, that there were advantages to be gained from using artificial mixtures of materials with one component fibrous, such as straw in clay, for bricks, or horse hair in plaster of paris for ceilings.

Recently, with the advent of cheap and strong glass fibres, and with the discovery of a number of new fibre-forming materials with better properties than anything available heretofore, the interest in fibre reinforced materials has increased rapidly, and is still accelerating. Fibre-reinforced polymers are replacing metals in a whole host of situations where load-carrying capacity is important. More efficient aircraft, turbine engines, and cars can be produced with fibre composites, and worthwhile new applications for these materials are being found almost every day.

In order to appreciate the potential benefits to be gained from fibre reinforced materials, however, it is necessary first to be aware of what can be done with more traditional materials. The introductory section of this text therefore starts with a review of the important properties required for load-bearing materials, and discusses traditional materials in this context. Then follows a brief statement of the basic ideas of isotropic elasticity theory, and its extension to non-isotropic cases of interest for fibre reinforcement.

1.1. Conventional Materials

The important properties for load-bearing materials fall naturally into two groups: mechanical and non-mechanical. Mechanical properties include stiffness, strength, ductility, hardness, toughness, fatigue, and creep. These will be discussed first. Non-mechanical properties, which will be discussed later, include density, temperature resistance, corrosion resistance (including stress corrosion and hydrogen embrittlement), and cost.

1.1.1. Stiffness

The stiffness of a material, or its resistance to reversible deformation under load, is a very important mechanical property. In order to characterize the effect of the load on the material, it is normally converted to a stress, that is the force per unit area of cross-section acted on by the load. Expressed in these terms, it is then an indication of the forces experienced by the individual atoms in the material as a result of applying the load. The response of the atoms is to change their positions slightly. Hence, under a stretching, or tensile stress, the atoms move apart in the direction of the force. The distance moved, divided by the original distance, is called the strain.

The movement of the atoms under the action of applied loads can be observed and measured using X-ray diffraction or neutron diffraction. The combined movement of the atoms constituting the whole specimen being stressed can usually be measured quite easily using a sensitive distance gauge.

A stiff material is one which deforms very little. Young's modulus is normally used as a measure of this, and is defined as the ratio of the tensile stress to the strain produced. The unit normally used for Young's modulus is the Pascal (Pa). (These units are described in Appendix B.) Other stiffness parameters will be discussed in the second part of this chapter.

Few materials are isotropic, and so measurements of Young's modulus made in different directions will give different results. These differences are particularly marked with single crystals of non-metals. However, polycrystalline metals, polymers, ceramics are generally sufficiently isotropic that they can be assigned a single value for the Young's modulus. Young's moduli range from about 1 TPa for a favourable direction in a diamond crystal through 100 GPa for a metal, 1 GPa for a polymer, and 10 MPa for a rubber. Table 1.1 gives the moduli of a representative selection for metals, ceramics, and polymers. The modulus of well-made specimens of crystalline elements or compounds is a measure of the deformability of the bonds between the atoms, and hence will be the same for different samples of the material. This is not quite the case with metal alloys, while with some polymers quite large variations between the modulus of different specimens of chemically similar materials can be found.

1.1.2. Strength

The strength of a material is the stress required to break it. It differs markedly from the modulus, in being determined as much by the method of manufacture and previous history of the specimen, as by the nature of the atoms and their arrangement. Pure, annealed, metals, and non-fibrous polymers are weak in both tension and compression. Ceramics and hard materials are generally much weaker in tension than in compression. Some polymer fibres are very strong in tension, while modern alloys, developed for use as structural members, are generally strong in both tension and compression. Typical tensile strengths are given in Table 1.1 for a variety of materials, excluding fibres and wires, some properties of which are given in Table 3.1.

TABLE 1.1. *Strength and Stiffness of Various Materials*

Material	Young's Modulus (GPa)	Tensile strength (MPa)
Metals:		
Aluminium (pure, annealed)	71	60
High-strength aluminium alloy		
(Al–Zn–Cu–Mg)	71	650
Iron (cast)	152	360
High-strength iron alloy (marag-		
ing steel)	212	2000
Magnesium alloy (Mg–Zn–Zr)	45	340
Titanium alloy		
(Ti–Zn–Al–Mo–Si)	120	1400
Tungsten	411	1800
Zirconium alloy		
(zircaloy 2)	97	590
Ceramics:		
Alumina (high density)	400	280
Concrete	50	3.5
Glass (sheet)	70	70
Polymers:		
Epoxy resin	2.5	60
Polycarbonate	2.5	65
Polyethylene (branched)	0.2	10
Rubber (natural)	0.018	32
Rubber (fluorocarbon)	0.002	7
Wood (Douglas fir)	14	34†
Wood (white pine)	7.6	16†

† Parallel to the grain; long-term strength is half this.

1.1.3. Ductility, Hardness, and Toughness

Figure 1.1 shows the stress–strain curves obtained with two different metals. Here the stress is calculated using the initial cross-section of the material, and the strain is calculated directly from the change in length of the test section.

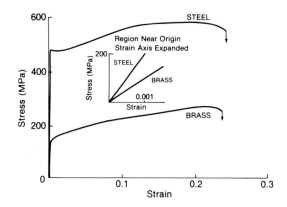

FIG. 1.1. Stress–strain curves for steel and brass.

With many metals the curve has three regions. First there is a linear region where the material is elastic, and on removal of the stress at any part of the line, the line is retraced back to the origin. The slope of the line gives the Young's modulus of the material.

The second part of the curve starts at the end of the linear region (this point is the yield point) and, except for a slight fall at the yield point in the case of some steels, the stress increases monotonically up to a maximum value, called the ultimate tensile strength.

In the third region the stress decreases monotonically with increasing strain until final failure occurs. In this region the specimen is not deforming uniformly; a region of reduced cross-section is formed (necking is occurring) and failure occurs at the neck.

(The maximum stress at the neck can be considerably greater than the stress in the rest of the material, so that the true stress at failure is much greater than that indicated by the stress axis. These two failure stresses are sometimes distinguished by calling the lower stress the engineering stress at failure, and the higher stress the true stress at failure. The true stress–true strain curve does not have a peak; the true stress increases monotonically to the failure point.)

In the second and third regions of the stress–strain curve, removal of the stress does not result in the material retracing the curve back to the origin. Instead, with decreasing stress the strain decreases at a rate governed by the modulus of the material, i.e. the line has a slope which is the same as that near the origin (Fig. 1.2). When the stress has been reduced to zero, there is now still some strain remaining. This strain is the origin of ductility. A stress–strain curve which is entirely linear up to the breaking-point, such as that obtained with a hard steel (e.g. razor blade steel) indicates very little ductility. A curve with a very long region after the yield point indicates a material with great ductility. With pure metals, the length of the specimen can sometimes be doubled before failure occurs.

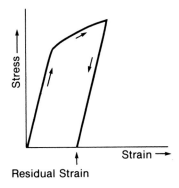

FIG. 1.2. Residual strain in a metal.

Figure 1.3 shows the stress–strain curves for two polymers. It can be seen that they show similar features to those observed with metals, except that yielding is generally not so sharp, and the curve does not usually have a monotonically decreasing region. Some polymers (e.g. polyesters) have very little ductility. Note that the yield point, or elastic limit, occurs at much higher strains for plastics than it does for most metals.

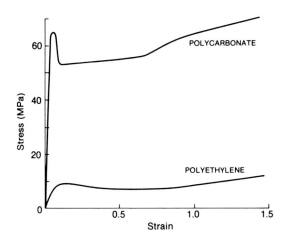

FIG. 1.3. Stress–strain curves for polyethylene and polycarbonate.

Hardness is a measure of the resistance of materials to plastic deformation, and hence ductile materials are usually not hard. It is determined by indenting the material with a hard ball, or a diamond with a pyramid-shaped tip, and measuring the size of indentation for a given load. Diamonds will leave indentations in glass, showing that plastic deformation, albeit small, is possible with glass. Table 1.2 gives some typical hardness values. They can be converted to roughly equivalent yield stresses in MPa by multiplying by three. Note, however, that with any metal or alloy, a wide range of hardness values can usually be obtained by heat treatment and cold working.

Toughness will be discussed in more detail later. It is often confused with ductility, and often erroneously equated with the area under the stress–strain curve. The measurement of toughness was put on a scientific footing with the advent of fracture mechanics. Fracture toughness or, more precisely, the work of fracture, is a measure of the work required to propagate a crack through unit area of material. Care is required in its measurement because, if a blunt crack is propagated into a specimen, an unrealistically high value for toughness is obtained. In addition, the impact methods of measuring toughness can give misleading results owing to the inclusion, in the result, of energy that has been dissipated as a result of processes unrelated to crack propagation. An example is the energy required to give the broken parts of the specimen some kinetic energy, so that they fly apart after fracture.

There is some connection between ductility and toughness, since ductile materials are generally tough, and hard materials are generally brittle. Table 1.2 gives some data on the work of fracture of some materials. The toughness of metals and alloys depends a great deal on heat treatment and work hardening, just as the hardness does. Thus toughness and hardness values are given for the metals and alloys in the same state.

TABLE 1.2. *Approximate Values for the Hardness and Toughness of Some Materials*

Material	Hardness (Kg/mm^2)	Work of fracture $(kJ\,m^2)$
Diamond	8400	–
Alumina	2600	0.02
Maraging steel	600	50
Hard tungsten	450	0.02
Glass	400	0.005
Titanium (99 %)	200	30
High-strength aluminium alloy	180	10
Pure aluminium	20	>100
Polycarbonate	20	3
Epoxy resin	18	0.1
Wood, across the grain	6	20
Wood, parallel to the grain	3	0.015

1.1.4. Fatigue

Materials which are subject to alternating stresses sometimes fail at stresses which are surprisingly low compared with their failure strength when tested under conditions of monotonically increasing stress. The seriousness of this problem was not fully appreciated until the disastrous failures of the first commercial jet passenger plane, the British Comet. This was made of what was considered to be an adequately strong aluminium alloy, yet the fuselage broke in two in mid-air. The failure started at the tip of a window, a crack propagating slowly therefrom under alternating loads, until it was so long that the material left was insufficient to support the load, so that complete failure of the whole fuselage suddenly occurred.

Nowadays materials are routinely tested for their fatigue resistance. The stressing regime used can be quite complicated, if it is desired to simulate service conditions. However, a good idea of the fatigue resistance can be obtained from a relatively simple test in which a piece of material is subjected to a sinusoidaly varying stress (Fig. 1.4). The stress amplitude required to break the piece is plotted as a function of the logarithm of the number of cycles it can withstand at that stress amplitude. When this test is carried out with aluminium alloys it is found that the breaking-stress decreases continuously as the number of cycles increases. Thus the material seems to have no strength after an infinite number of cycles. Iron alloys, on the other hand, still retain about half their strength after a very large number of cycles. The difference between the two is illustrated in Fig. 1.4. The fatigue or endurance limit is the stress amplitude of the plateau region observed after a large number of cycles.

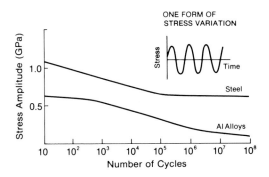

FIG. 1.4. Fatigue curves for aluminium and steel.

Fatigue failure starts with plastic deformation at the surface of the specimen. Then very fine cracks begin to appear, and one of these gradually increases in length until it becomes big enough for the material to fail by the normal fracture process. The number of cycles to failure can vary greatly from specimen to specimen of the same material, even when great pains have been taken to make them identical.

1.1.5. Creep

Materials under constant stress can gradually extend by plastic deformation. The stress to cause this creep can be considerably less than the tensile strength. With metals the process is insignificant at temperatures less than about half the absolute melting-temperature. Thus the creep of the metals normally used for bridges and aircraft (steel and aluminium), for example, is negligible at room temperature. Polymers, however, can creep significantly at room temperature.

Figure 1.5 shows a typical creep curve obtained with an aluminium alloy. It may be seen that the curve has three regions. The first, primary creep region, has a relatively high creep rate. This decreases with time until a constant creep rate is achieved in the secondary creep region. Eventually the specimen has deformed to such an extent that necking starts, and the creep rate increases once again. This is the tertiary creep region, and it generally continues for a short period only, whereupon the specimen fails.

Polymers show similar behaviour (Fig. 1.6).

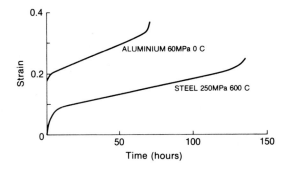

FIG. 1.5. Creep curves for aluminium and steel.

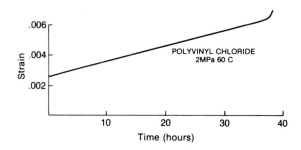

FIG. 1.6. Creep curve for polyvinyl chloride.

1.1.6. Density

A non-mechanical property of great importance is density. High density is an advantage for materials used for hammers, road rollers, flywheels, etc. These are exceptional cases, however. Generally, light materials have significant advantages. Aircraft and spacecraft are obvious examples of structures which need light materials, but any rapidly rotating part not used as an energy store is best made of light materials, since centrifugal forces are thereby reduced. Such parts as turbine rotors, for example, should be made of materials which are as light as possible, taking into account strength and deformation also. Cars used in cities would be more efficient, if lighter. In addition, since a large part of the strength of the members supporting bridges is used up sustaining the dead weight of the bridge structure, there would be benefits from using light materials for bridges, and other static load-bearing structures, again taking into account strength and stiffness as well.

In many applications, the density criterion should be combined with strength and stiffness, so that an efficiency criterion is obtained. For strength, the most obvious criterion is that the strength/density ratio should be as great as possible. Similarly for stiffness, the obvious criterion is that the modulus/density ratio should be as large as possible.

However, when account is taken of strength and stiffness, the role of density is not as great as might be expected. With metals, it is found that the modulus, which governs stiffness, and also to some extent controls the strength attainable (this will be discussed in the next chapter), is roughly proportional to the density, i.e. the modulus/density ratio is approximately constant. In the case of polymers the modulus, and hence stiffness, is far too low for their use in most structures, unless the polymer is reinforced. Thus the low density of unreinforced polymers has little significance for load-bearing structures which have to retain nearly their original shape when loaded. Ceramic materials show little relation between modulus and density. However, their extreme brittleness makes them unsafe in tensile load-bearing applications, so they are only used in special circumstances, when some other property, such as resistance to high temperature, is of overriding importance.

Why, then, is aluminium used for subsonic aircraft, and wood used for gliders? The reason is that many of the parts have to resist bending or buckling. Aircraft wings can only withstand a deformation under load of a few hundredths of a per cent before they lose the shape required to keep the craft aloft efficiently. Figure 1.7 shows what an aircraft would look like with 1.6% strain in the wing spar booms.

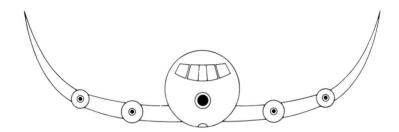

F<small>IG</small>. 1.7. Plane with strain of 0.016 in wing structure. (After Gordon, J. E. (1964) *Proc. Roy. Soc.* **A282,** 16.)

In the bending of a beam, the deformation results in curvature, of radius R, say. If the moment of forces causing the bending is M, then the basic formula for bending is

$$MR = EI \qquad (1.1)$$

where E is the Young's Modulus of the beam and I is the moment of area of the beam, which for a beam of thickness d is $d^3/12$ per unit width. A beam with great resistance to bending has a high value of MR, and hence of EI, or $Ed^3/12$ (often called the flexural rigidity). The weight of beam per unit length and width is ρd. Now we can compare the weights of two materials having the same flexural rigidity. We will distinguish the materials by the subscripts 1 and 2. For the same flexural rigidity, they will have different thicknesses, related by the expression

$$E_1 d_1^3 = E_2 d_2^3 \qquad (1.2)$$

The ratio of their weights will be

$$\frac{w_1}{w_2} = \frac{\rho_1 d_1}{\rho_1 d_2} \qquad (1.3)$$

and using equation (1.2) this comes to

$$\frac{w_1}{w_2} = \frac{\rho_1/E_1^{1/3}}{\rho_2/E_2^{1/3}} \qquad (1.4)$$

Clearly the most efficient material has the largest value of $E^{1/3}/\rho$.

The same result is obtained when we consider compressive failure of long thin beams, such as are used, for example, in aeroplane wings. These beams fail by buckling, and the compressive failure load is governed by the flexural rigidity.

Table 1.3 compares a number of structural materials on the basis of E/ρ and $E^{1/3}/\rho$. It will be seen that for most metals the values for E/ρ are close to 25, apart from zirconium. For wood the value is somewhat higher, while for polyethylene (and almost all other man-made polymers) the value is very much less. However, the values for $E^{1/3}/\rho$ for the lighter materials such as wood, aluminium, and magnesium are much larger than for the heavier materials. Wood comes out best, and hence its use for gliders. Aluminium alloys provide a good compromise for strength as well as $E^{1/3}/\rho$, and are more durable than wood, and hence are used for subsonic aircraft. (Some excellent small planes have been made with wood, however.)

TABLE 1.3. *Stiffness-Density Parameters for Some Materials*

Material	Density ρ (Mg m^{-3})	Young's Modulus, E (GPa)	E/ρ (MmN kg^{-1})	$E^{1/3}/\rho$
Aluminium	2.70	71	26.3	1.53
Magnesium	1.74	42	24.1	2.04
Polyethylene	0.93	0.2	0.22	0.63
Steel	7.87	212	26.9	0.76
Titanium	4.51	120	26.6	1.09
Tungsten	19.3	411	21.3	0.39
Wood (Sitka spruce)	0.39	13	33.3	6.03
Zirconium	6.49	94	14.5	0.70

1.1.7. Temperature Resistance

Another important non-mechanical property is resistance to extremes of temperature.

Polymers generally have a small range of temperatures over which they can be used. They tend to become brittle when cold, and soft when hot. Some nylons start to soften at as little as 60 C (C is used throughout for Celsius).

Brittleness in polymers generally occurs at temperatures below the glass transition temperature, when the polymer changes from a viscous liquid to a glass-like solid. For most amorphous polymers this occurs between + 50 and − 50 C. Even rubbers become brittle below about − 70 C. The onset of brittleness means that these materials should be used with great care for tensile or flexural loads at low temperatures.

At the upper end of their temperature range polymers soften, then melt or decompose (or both). Even the best high temperature polymers (e.g. polyimides) do not last very long at 400 C. Table 1.4 lists some typical thermal properties of some polymers.

Some steels also become brittle below 0 C and until this danger was widely recognized, disastrous failures of large steel structures such as ships, bridges, and pressure vessels were

TABLE 1.4. *Some Thermal Properties of Polymers*

Polymer	Temperature (C)		
	Glass transition	Softening	Melting
Polyethylene (linear)	−78	50	140
Polycarbonate	160	135	270
Epoxy	†	200	†
Polyester	†	150	†

† Values not given because these polymers do not melt, and their glass transition temperatures can vary from 40 C to 200 C according to composition and state of cure

not uncommon. Crystalline phase changes can also cause problems. The most striking case of this is tin, a solid piece of which can go to a powder when taken through the phase change that occurs at 13 C.

The problems at low temperatures can normally be avoided by suitable choice of materials. Materials to withstand high temperatures are much harder to find. Not only is there the danger of the material softening, but it can also suffer acclerated corrosion, and creep rupture.

Conventional strong steels tend to lose their strength by changes in microstructure at temperatures in the range 200–300 C. Special steels for high temperature use have been developed; one of the more notable types are the maraging steels which are useful up to 450–500 C. At higher temperatures, chromium must be added to impart corrosion resistance, and this enables temperatures up to 800 C to be reached, although at lower strength levels (about 350 MPa). For higher temperatures than this, cobalt based alloys are used, and such an alloy will withstand a stress of 150 MPa at 930 C. At still higher temperatures nickel based alloys may be used, strength levels of 150 MPa being achieved for 100 hours at 1000 C.

Aluminium alloys are only suitable for relatively low temperatures, and even aluminium alloys developed for high temperature use have short-term strengths of only about 50 MPa at 400 C. When light weight is important, titanium alloys are used. These can have short-term strengths of about 400 MPa at 700 C.

For temperatures above 1100 C ceramics have to be used. The loads that these can carry, however, is very limited. This is because they are extremely brittle, and the risk of catastrophic failure when under load at temperatures below the normal operating temperature is very great.

The production of more efficient engines for aircraft and land-based applications requires the development of better high-temperature materials. A significant portion of the research on new materials is devoted to this problem.

1.1.8. Corrosion Resistance

Even at moderate temperatures corrosion can be a serious problem. A familiar example of this is the rusting of the bodies of cars driven on the salty slush encountered in some cities with a heavy snowfall. Corrosion is a problem which is now well understood. It can often be solved by suitable choice of materials, avoidance of the use of combinations of metals which set up unfavourable electrochemical reactions under moist conditions, the use of sacrificial electrodes, and the protection of corrodable surfaces by such materials as paints. Since corrosion can cause total loss of fair-sized regions of load-bearing members, consideration of the corrosion conditions likely to be encountered is necessary when a load-bearing structure is being designed.

High-temperature corrosion is a much more difficult problem, and provides an upper temperature limit to the usefulness of some metals and alloys. Protection of metal surfaces by coatings of various sorts is seldom a suitable remedy, except for use at moderate temperatures, since thermal expansion differences between the coating and the metal often results in fragmentation of the coating and spalling off, after a few cycles up to high temperature and back to room temperature. As mentioned in the previous section, if stainless steels, or nickel, chromium, and cobalt alloys are not sufficiently corrosion resistant for a high-temperature load-bearing structure, then ceramics normally have to be used, with their inherent disadvantages of very low load-bearing capacity.

When a material is stressed it is often more susceptible to corrosion than when unstressed. This phenomenon is called stress corrosion and is due to the greater chemical reactivity of a material when the atoms are not in their equilibrium positions in the structure. An allied phenomenon is that of hydrogen embrittlement. Hydrogen, often produced by corrosive attack, can enter the material as single atoms, and there form weak hydrides (e.g. in Zr), or collect in voids in the form of molecular hydrogen and weaken the material by exerting high pressures. Materials which are subject to these problems have to be provided with protective layers, if alternative materials are not available.

1.1.9. Cost

Probably the most important characteristic of a material is its cost. However, direct cost, on a weight or volume basis (i.e. dollars per kilogram or dollars per cubic metre) is seldom the operative criterion. In the building industry such a criterion may be important, since the most heavily used materials—steel, wood, bricks, plaster, and concrete, are the cheapest. They range in price from about $0.01–0.1/kg for bricks, plaster, and concrete, to about $0.1–1/kg for steel and wood. This clearly is not true, though, in aircraft, which require the use of relatively expensive materials such as aluminium. The more comprehensive concept of cost effectiveness is relevant in this case, and indeed for most uses of materials. For example, the use of the very expensive carbon fibre reinforced panels in the Boeing 747 "Jumbo Jet" is worth while because of the financial benefits that accrue from the resultant increase in the load-carrying capacity, and hence profitability, of the plane.

Cost effectiveness is a criterion which depends very greatly on the end use of the structure containing the material, and generalizations are difficult to make. It usually involves such factors as: cost per unit load supported, cost per unit of deformation permitted under load, fabrication cost, and general efficiency.

1.2. Elasticity Theory

In this section elasticity theory will be briefly introduced and the terminology used later for stresses and strains will be described. Two classes of stress are normally considered.

A body being pulled at each end by a force F (Fig. 1.8) is in uniaxial tension, and the TENSILE STRESS, σ, is equal to F, divided by the area over which it acts, i.e. the area of cross-section of the bar. The bar will become longer due to the stress, and the strain, ε, is the change in length divided by the original length, i.e. $\delta l/l$.

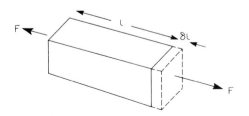

FIG. 1.8. Rod in uniaxial tension.

Figure 1.9 shows a rectangular body being sheared by a force F. The SHEAR STRESS, τ, is equal to F, divided by the area of which it acts, i.e. the area of the top of rectangular box. The box will deform due to the stress, and the change in angle of the corners of the box γ, is the shear strain.

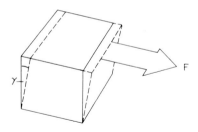

FIG. 1.9. Simple shear.

1.2.1. Stresses

We will now analyse bodies subject to more general stresses. Let us consider a small element of a solid body under stress. Any place in the body can be defined by cartesian axes x, y, and z. Suppose the origin of these axes is at one corner of the small element of the body (Fig. 1.10). For simplicity we will consider a rectangular shaped element, whose sides have lengths dx, dy, and dz as shown. The stresses in the body can be resolved into components in the three directions. Thus we will have tensile stresses, σ_x, acting on the lower left face, and, in the opposite direction, on the opposite face of the body (not shown). Similarly there will be equal and opposite stresses, σ_y, acting on the faces normal to the y axis, and σ_z acting on the faces normal to the z axis.

The forces acting on each face in these directions are the product of the stress and the area of the face. For example for the faces normal to the x axis they are $\sigma_x dy dz$.

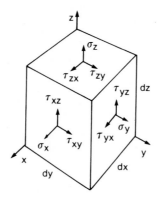

FIG. 1.10. Stresses on an elementary volume.

Negative values for σ_x, σ_y, σ_z indicate compressive stresses.

In addition to the tensile stresses there will, in general, be shear stresses. The symbol τ will be used for the shear stress and the subscripts indicate the plane on which it acts, and the direction. Thus τ_{xy} is the shear in the plane normal to the x axis, acting in the y direction. (The first subscript is for the plane, the second for the direction.) Shears also act on the opposite faces to those shown, and, again are in the opposite directions.

Although the subscripts suggest that there are six shears, in fact there are only three. This is apparent when we examine the tendency of the element to rotate under the action of the shears.

Consider the top and bottom faces of the element, together with faces normal to the y axis. Figure 1.11 shows the shears tending to rotate this.

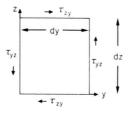

FIG. 1.11. Pure shear.

We will allow the dimensions of the element dz, dy (and dx) to go to zero, so that we can neglect body forces. (Body forces are forces that are not caused by externally applied loads. An example of a body force is the force arising from the action of the earth's gravitational field on the mass of the element.) The clockwise moment is given by the forces at top and bottom (Fig. 1.11), $\tau_{zy}\,dx\,dy$, multiplied by their distance apart, dz, i.e. $\tau_{zy}\,dx\,dy\,dz$. Similarly, the anticlockwise moment is $\tau_{yz}\,dx\,dz$ (the forces) multiplied by dy (the distance). For equilibrium (i.e. no rotation) the clockwise and anticlockwise moments must be equal:

$$\tau_{zy}dx\,dy\,dz = \tau_{yz}\,dx\,dz\,dy$$

Thus $\tau_{zy} = \tau_{yz}$. Similarly $\tau_{xy} = \tau_{yx}$ and $\tau_{xz} = \tau_{zx}$.

We thus conclude that there are six components of stress: three tensile stresses, σ_x, σ_y and σ_z, and three shears τ_{xy}, τ_{yz}, τ_{xz}.

Only a limited number of problems have been solved when all six stresses have significant values, and all possible deformations resulting therefrom have to be considered. Fortunately, many systems reduce to the relatively simple planar form. For example, a very common problem is the bending of beams under a variety of loading systems. Very often it is possible to neglect the stresses in the direction of the thickness of the beam (the z direction for the orientation of axes shown in Fig. 1.12). This system is called PLANE STRESS, and in the case shown we can assume $\sigma_z \simeq \tau_{xz} \simeq \tau_{yz} \simeq 0$. Stress analysis shows that the other stresses in the centrally loaded beam shown in Fig. 1.12 depend on the distance from the centre, as follows:

$$\sigma_x = - F\,(l-x)y/I$$
$$\sigma_y = 0$$
$$\tau_{xy} = F\,(d^2 - 4y^2)/8I$$

where
$$I = bd^3/12$$

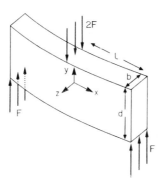

FIG. 1.12. Plane stress flexure.

1.2.2. Resolution of Stresses

Stresses acting on any plane can be calculated if the values of the six stresses are all known. For simplicity we will consider plane stress, with σ_z, τ_{yz} and τ_{xz} equal to zero. We will consider a small element of material of unit thickness in the z direction, with sides dx and dy, as shown in Fig. 1.13, so that $\tan \phi = dx/dy$. If we divide the element in two along the diagonal, the halves will only stay in position if held together by a stress σ and prevented from sliding by a shear τ. The values of σ and τ can be related to the stresses σ_x, σ_y, and τ_{xy}.

Let the forces acting across the plane be F_x and F_y (see Fig. 1.13). Equating forces in the x and y directions.

$$F_x = \sigma_x\,dy - \tau_{xy}\,dx \tag{1.5}$$

$$F_y = \sigma_y\,dx - \tau_{xy}\,dy \tag{1.6}$$

We can now resolve F_x and F_y along the normal to the plane to find σ. Thus

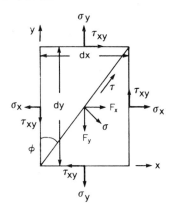

FIG. 1.13. Stresses and forces acting on a diagonal plane.

$$\sigma\sqrt{(dx^2 + dy^2)} = F_x \cos\phi + F_y \sin\phi \qquad (1.7)$$

and using the equations for F_x and F_y ((1.5) and (1.6)) and noticing that $\cos\phi = dy/\sqrt{(dx^2 + dy^2)}$ and $\sin\phi = dx/\sqrt{(dx^2 + dy^2)}$ we find that

$$\sigma = \sigma_x \cos^2\phi + \sigma_y \sin^2\phi - 2\tau_{xy} \sin\phi \cos\phi \qquad (1.8)$$

Resolving forces in the plane we obtain

$$\tau\sqrt{(dx^2 + dy^2)} = F_x \sin\phi - F_y \cos\phi \qquad (1.9)$$

and making the same substitutions yields

$$\tau = (\sigma_x - \sigma_y) \sin\phi \cos\phi + \tau_{xy} (\cos^2\phi - \sin^2\phi) \qquad (1.10)$$

We can choose ϕ so that $\tau = 0$. For this

$$\frac{\tau_{xy}}{\sigma_y - \sigma_x} = \frac{\sin\phi \cos\phi}{\cos^2\phi - \sin^2\phi} = \tfrac{1}{2}\tan 2\phi \qquad (1.11)$$

This equation shows that the shears are zero in two directions. When the x and y axes are in these directions, they are called PRINCIPAL AXES.

A very important result of equation (1.10) is that when a piece of material is subject to uniaxial tension, e.g. $\tau_{xy} = \sigma_x = 0$ (so that σ_y is the only stress) then

$$\tau = \sigma_y \sin\phi \cos\phi \qquad (1.12)$$

This shear has its maximum value when $\phi = \pi/4$, and is then $\tau = \sigma_y/2$. Thus, in a tensile test, very large shear forces can be produced, and ductile materials tested in this way normally fail by shearing at about 45° to the tensile axis.

In the beam shown in Fig. 1.12 τ_{xy} has its maximum value in the centre plane ($y = 0$). However, the maximum shear stresses are at the upper and lower surfaces under the load, and have the (absolute) value

$$\tau_{max} = Fld/2I. \qquad (1.13)$$

1.2.3. Strains

In the same way that six stresses are needed to describe the state of stress of a body, six strains are needed to represent the state of deformation. The strains normally considered in elasticity theory are small, usually less than 0.001. It would be impractical to show such small strains in a diagram, so Fig. 1.14 shows a much exaggerated view of strains in the x–y plane. We again have an element with sides which were originally of lengths dx (OA) and dy (OB). As a result of the deformation O has moved to O', A to A' and B to B'. What was originally a right angle at O has become an angle $\pi/2 - \theta - \beta$, and the lengths OA and OB have both increased.

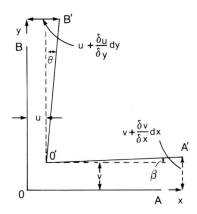

FIG. 1.14. Deformation of an L-shaped element.

If the displacements of O in the x and y directions are u and v respectively, then the corresponding displacements for A in these directions are $u + \dfrac{\partial u}{\partial x} dx$ and $v + \dfrac{\partial v}{\partial x} dx$ respectively. For B they are $v + \dfrac{\partial v}{\partial y} dy$ and $u + \dfrac{\partial u}{\partial y} dy$ respectively.

The change in length of OA (assuming $\theta \simeq 0$) is the displacement of A in the x direction less the displacement of O in the x direction i.e. $u + \dfrac{\partial u}{\partial x} dx - u$. Thus the tensile strain ($=$ change in length/original length) is $\partial u/\partial x$. Similarly, the strain in the y direction is $\partial v/\partial y$ and in the z direction (not shown) is $\partial w/\partial z$, for a displacement of O in the z direction of w. If we write ε_x, ε_y, and ε_z for these strains, we therefore have

$$\varepsilon_x = \partial u/\partial x \tag{1.14}$$

$$\varepsilon_y = \partial v/\partial y \tag{1.15}$$

$$\varepsilon_z = \partial w/\partial z \tag{1.16}$$

Compressive strains will be shown by negative values for ε_x, ε_y, or ε_z.

The rectangular element has also been sheared, so that the angle at O has decreased by $\theta + \beta$. In Fig. 1.14 it can immediately be seen that, for θ and β sufficiently small, we can write $\theta = \tan \theta$ and $\beta = \tan \beta$, so that $\theta = \partial u/\partial y$ and $\beta = \partial v/\partial x$. The total shear strain is $\theta + \beta$, so writing γ_{xy} for this, we have

$$\gamma_{xy} = \partial v/\partial x + \partial u/\partial y \tag{1.17}$$

and similarly

$$\gamma_{yz} = \partial w/\partial y + \partial v/\partial z \tag{1.18}$$

$$\gamma_{xz} = \partial w/\partial x + \partial u/\partial z \tag{1.19}$$

Plane strain, i.e. when all the strains normal to some plane in the body are zero, is analogous to plane stress. It is quite commonly assumed for the treatment of a stress problem, and often permits of relatively simple analysis.

1.2.4. Elastic constants

Only two constants are needed to describe the elastic behaviour of an isotropic material. For convenience, however, four are in common use. Each can be calculated if two of the others are known.

The most readily available elastic constant is Young's modulus, values of which have been given in Table 1.1. This is measured in a uniaxial tensile test, e.g. σ_x has a value, but all other stresses are zero. It is defined by an equation of the type

$$E = \sigma_x/\varepsilon_x \tag{1.20}$$

In a tensile test it is found that contractions occur normal to the direction of the applied stress. These may be used to determine another elastic constant, Poisson's ratio, v, i.e.

$$v = -\varepsilon_y/\varepsilon_x = -\varepsilon_z/\varepsilon_x \tag{1.21}$$

for uniaxial tension in the x direction.

Since we only consider small stresses and strains, we can superpose stresses and strains arising from different applied loads by simple addition. Thus, for a system where $\sigma_x, \sigma_y, \sigma_z,$ $\tau_{xy}, \tau_{yz},$ and τ_{zx} may all have values we can write, for the total tensile strains in each direction,

$$\varepsilon_x = \frac{1}{E}\{\sigma_x - v(\sigma_y + \sigma_z)\} \tag{1.22}$$

$$\varepsilon_y = \frac{1}{E}\{\sigma_y - v(\sigma_x + \sigma_z)\} \tag{1.23}$$

$$\varepsilon_z = \frac{1}{E}\{\sigma_z - v(\sigma_x + \sigma_y)\}. \tag{1.24}$$

For most metals v is close to 0.3. However, for rubbers it is almost exactly 0.5 (this indicates no volume change when stressed in uniaxial tension) while for glass it is about 0.22.

The third elastic constant to be considered is the shear modulus, G, defined by an expression of the type,

$$G = \tau_{xy}/\gamma_{xy} \tag{1.25}$$

(i.e. shear modulus = shear stress/shear strain).

If we consider an element of a body of unit length in the z direction, subject to a tension in the y direction, and an equal but opposite compression in the x direction, and all other stresses zero, we can derive the relation between the shear modulus and the two preceding elastic constants. This state of stress is called PURE SHEAR. The body is deformed without any rotation. In contrast, a body sheared by a force parallel to one surface suffers deformation and rotation; this is called SIMPLE SHEAR, and is illustrated in Fig. 1.9.

For our pure shear case we can write $\sigma_y = -\sigma_x = \sigma$, say.

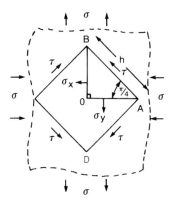

FIG. 1.15. Square element in pure shear.

We will consider a 45° isoceles triangle OAB, within our element of material, Fig. 1.15. Let the hypotenuse have a length h. The forces acting on the triangle are as shown in Fig. 1.15. No stress has been shown acting across the hypotenuse. Suppose this were σ^1, then equating forces normal to the hypotenuse we have,

$$h\sigma^1 = h\cos\frac{\pi}{4}.\sigma_x\cos\frac{\pi}{4} + h\cos\frac{\pi}{4}.\sigma_y\cos\frac{\pi}{4} \tag{1.26}$$

or

$$\sigma^1 = (\sigma_x + \sigma_y)\cos^2\frac{\pi}{4} = 0 \tag{1.27}$$

i.e. there is no stress acting across AB.
Equating forces parallel to the hypotenuse we obtain

$$h\tau = h\cos\frac{\pi}{4}.\sigma_y\cos\frac{\pi}{4} - h\cos\frac{\pi}{4}.\sigma_x\cos\frac{\pi}{4} \tag{1.28}$$

or

$$\tau = (\sigma_y - \sigma_x)\cos^2\frac{\pi}{4} = \sigma. \tag{1.29}$$

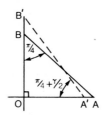

FIG. 1.16. Deformation of a sheared element.

Now consider the deformation. OA will be shortened to OA', and OB lengthened to OB', Fig. 1.16. The angle $O\hat{A}B$ will have changed by an amount $\gamma/2$, so that $D\hat{A}B$, which was a right angle (Fig. 1.15), will have changed to $\pi/2 + \gamma$; i.e. γ is the total shear strain for the element. The relative changes in the lengths OA and OB are given by the strains ε_x and ε_y respectively, where, using equations (1.22) and (1.23).

$$\varepsilon_x = (\sigma_x - v\sigma_y)/E = -(1+v)\sigma/E \qquad (1.30)$$

and

$$\varepsilon_y = (\sigma_y - v\sigma_x)/E = (1+v)\sigma/E \qquad (1.31)$$

Now $\tan O\hat{A}B'$ is OB'/OA'; thus

$$\tan\left(\frac{\pi}{4} + \frac{\gamma}{2}\right) = \frac{OB'}{OA'} = \frac{1+\varepsilon_y}{1+\varepsilon_x} = \frac{1+(1+v)\sigma/E}{1-(1+v)\sigma/E} \qquad (1.32)$$

(remembering that ε_x is negative and using equations (1.30) and (1.31)). The tangent can also be expanded; thus for small γ

$$\tan\left(\frac{\pi}{4} + \frac{\gamma}{2}\right) = \frac{\sin\left(\frac{\pi}{4} + \frac{\gamma}{2}\right)}{\cos\left(\frac{\pi}{4} + \frac{\gamma}{2}\right)} \simeq \frac{\sin\frac{\pi}{4} + \frac{\gamma}{2}\cos\frac{\pi}{4}}{\cos\frac{\pi}{4} - \frac{\gamma}{2}\sin\frac{\pi}{4}} \qquad (1.33)$$

therefore

$$\tan\left(\frac{\pi}{4} + \frac{\gamma}{2}\right) \simeq \frac{1 + \gamma/2}{1 - \gamma/2} \qquad (1.34)$$

Inspecting equations (1.32) and (1.34) we can see that

$$\gamma = 2(1+v)\sigma/E \qquad (1.35)$$

We now have both τ and γ in terms of the stress, σ, and the elastic constants. Thus, using equations (1.29) and (1.35), we have for the shear modulus, G

$$G = \frac{\tau}{\gamma} = \frac{E}{2(1+v)} \qquad (1.36)$$

The fourth elastic constant to be considered is the resistance of the material to uniform compression, called the bulk modulus. The symbol K will be used for this. The modulus is defined by the expression,

$$K = -P/e \tag{1.37}$$

where P is the pressure, which acts on all faces of an element of material, and e is the change in volume per unit of volume. Let the element of material be a cube, with a pressure P acting on each face. If the side of the cube initially is l, and if the sides of the cube are parallel to the axes (Fig. 1.17), then the sides will be reduced to lengths $l(1 + \varepsilon_x)$, $l(1 + \varepsilon_y)$ and $l(1 + \varepsilon_z)$ where

$$\varepsilon_x = \varepsilon_y = \varepsilon_z = -P(1 - 2v)/E \tag{1.38}$$

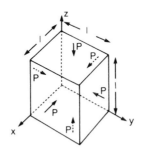

FIG. 1.17. Triaxial compression.

(using equations (1.22), (1.23), and (1.24) with $\sigma_x = \sigma_y = \sigma_z = -P$). The new volume of the cube is

$$(1 + e)l^3 = (1 + \varepsilon_x)(1 + \varepsilon_y)(1 + \varepsilon_z)l^3 \tag{1.39}$$

Expanding the right hand side of equation (1.39) and neglecting second and third orders products (e.g. $\varepsilon_x \varepsilon_y$ etc., and $\varepsilon_x \varepsilon_y \varepsilon_z$) and simplifying we find that the equation reduces to,

$$e = \varepsilon_x + \varepsilon_y + \varepsilon_z \tag{1.40}$$

Using equation (1.38) therefore, we find that

$$e = -3P(1 - 2v)/E \tag{1.41}$$

so that the bulk modulus comes to

$$K = -\frac{P}{e} = \frac{E}{3(1 - 2v)} \tag{1.42}$$

We have thus derived expressions for the shear modulus and the bulk modulus in terms of Young's modulus and Poisson's ratio.

1.2.5. Anisotropic Materials

It was mentioned earlier that single crystals of materials are usually anisotropic. Fibre reinforced materials are often highly anisotropic. While two elastic constants are required to describe the elastic properties of isotropic materials, twenty-one are required for anisotropic materials with no planes of symmetry.

As with isotropic materials, there are normally six independent stresses and six independent strains for anisotropic materials. The tensile stresses are expressed thus: σ_1, σ_2, σ_3, while the shear stresses are expressed thus: τ_{23}, τ_{31}, τ_{12}, and the strains are $\varepsilon_1 = \partial u/\partial x$ etc., and $\gamma_{23} = \partial v/\partial z + \partial w/\partial y$ etc. (i.e. the subscripts x, y, z used for isotropic materials are replaced by 1, 2, and 3). The stress–strain relations are best expressed in matrix form:

$$
\begin{vmatrix} \sigma_1 \\ \sigma_2 \\ \sigma_3 \\ \tau_{23} \\ \tau_{31} \\ \tau_{12} \end{vmatrix} = \begin{vmatrix} C_{11} & C_{12} & C_{13} & C_{14} & C_{15} & C_{16} \\ C_{12} & C_{22} & C_{23} & C_{24} & C_{25} & C_{26} \\ C_{13} & C_{23} & C_{33} & C_{34} & C_{35} & C_{36} \\ C_{14} & C_{24} & C_{34} & C_{44} & C_{45} & C_{46} \\ C_{15} & C_{25} & C_{35} & C_{45} & C_{55} & C_{56} \\ C_{16} & C_{26} & C_{36} & C_{46} & C_{56} & C_{66} \end{vmatrix} \cdot \begin{vmatrix} \varepsilon_1 \\ \varepsilon_2 \\ \varepsilon_3 \\ \gamma_{23} \\ \gamma_{31} \\ \gamma_{12} \end{vmatrix} \qquad (1.43)
$$

The $[C_{ij}]$ matrix is the stiffness matrix. If it is required to calculate strains the compliance matrix, $[S_{ij}]$ is used:

$$
\begin{vmatrix} \varepsilon_1 \\ \varepsilon_2 \\ \varepsilon_3 \\ \gamma_{23} \\ \gamma_{31} \\ \gamma_{12} \end{vmatrix} = \begin{vmatrix} S_{11} & S_{12} & S_{13} & S_{14} & S_{15} & S_{16} \\ S_{12} & S_{22} & S_{23} & S_{24} & S_{25} & S_{26} \\ S_{13} & S_{23} & S_{33} & S_{34} & S_{35} & S_{36} \\ S_{14} & S_{24} & S_{34} & S_{44} & S_{45} & S_{46} \\ S_{15} & S_{25} & S_{35} & S_{45} & S_{55} & S_{56} \\ S_{16} & S_{26} & S_{36} & S_{46} & S_{56} & S_{66} \end{vmatrix} \cdot \begin{vmatrix} \sigma_1 \\ \sigma_2 \\ \sigma_3 \\ \tau_{23} \\ \tau_{31} \\ \tau_{12} \end{vmatrix} \qquad (1.44)
$$

Notice that both these matrices are symmetric, i.e. $C_{ij} = C_{ji}$ and $S_{ij} = S_{ji}$

The number of independent elastic constants is only nine for orthotropic materials. Laminae of fibre reinforced materials, with all the fibres parallel, are examples of orthotropic materials. These materials have three mutually orthogonal planes of symmetry for mechanical properties, and there are no interactions between normal stresses, σ_1, σ_2, and σ_3 and shearing strains, γ_{23}, γ_{31}, γ_{12}. Conversely there are no interactions between shearing stresses and normal strains. Thus equation (1.43) reduces to

$$
\begin{vmatrix} \sigma_1 \\ \sigma_2 \\ \sigma_3 \\ \tau_{23} \\ \tau_{31} \\ \tau_{12} \end{vmatrix} = \begin{vmatrix} C_{11} & C_{12} & C_{13} & 0 & 0 & 0 \\ C_{12} & C_{22} & C_{23} & 0 & 0 & 0 \\ C_{13} & C_{23} & C_{33} & 0 & 0 & 0 \\ 0 & 0 & 0 & C_{44} & 0 & 0 \\ 0 & 0 & 0 & 0 & C_{55} & 0 \\ 0 & 0 & 0 & 0 & 0 & C_{66} \end{vmatrix} \cdot \begin{vmatrix} \varepsilon_1 \\ \varepsilon_2 \\ \varepsilon_3 \\ \gamma_{23} \\ \gamma_{31} \\ \gamma_{12} \end{vmatrix} \qquad (1.45)
$$

and equation (1.44) reduces to

$$
\begin{vmatrix} \varepsilon_1 \\ \varepsilon_2 \\ \varepsilon_3 \\ \gamma_{23} \\ \gamma_{31} \\ \gamma_{12} \end{vmatrix} = \begin{vmatrix} S_{11} & S_{12} & S_{13} & 0 & 0 & 0 \\ S_{12} & S_{22} & S_{23} & 0 & 0 & 0 \\ S_{13} & S_{23} & S_{33} & 0 & 0 & 0 \\ 0 & 0 & 0 & S_{44} & 0 & 0 \\ 0 & 0 & 0 & 0 & S_{55} & 0 \\ 0 & 0 & 0 & 0 & 0 & S_{66} \end{vmatrix} \cdot \begin{vmatrix} \sigma_1 \\ \sigma_2 \\ \sigma_3 \\ \tau_{23} \\ \tau_{31} \\ \tau_{12} \end{vmatrix} \qquad (1.46)
$$

for axes taken parallel to the intersections of the symmetry planes.

Some materials are transversely isotropic. This is true for sheets of randomly oriented fibre composites. These materials have mechanical properties which are equal in all directions in a plane, and have only five independent elastic constants. Taking the 1–2 plane as the plane of isotropy, equations (1.45) and (1.46) reduce to

$$
\begin{vmatrix} \sigma_1 \\ \sigma_2 \\ \sigma_3 \\ \tau_{23} \\ \tau_{31} \\ \tau_{12} \end{vmatrix} = \begin{vmatrix} C_{11} & C_{12} & C_{13} & 0 & 0 & 0 \\ C_{12} & C_{11} & C_{13} & 0 & 0 & 0 \\ C_{13} & C_{13} & C_{33} & 0 & 0 & 0 \\ 0 & 0 & 0 & C_{44} & 0 & 0 \\ 0 & 0 & 0 & 0 & C_{44} & 0 \\ 0 & 0 & 0 & 0 & 0 & C_{66} \end{vmatrix} \cdot \begin{vmatrix} \varepsilon_1 \\ \varepsilon_2 \\ \varepsilon_3 \\ \gamma_{23} \\ \gamma_{31} \\ \gamma_{12} \end{vmatrix} \tag{1.47}
$$

where $C_{66} = \frac{1}{2}(C_{11} - C_{12})$, and

$$
\begin{vmatrix} \varepsilon_1 \\ \varepsilon_2 \\ \varepsilon_3 \\ \gamma_{23} \\ \gamma_{31} \\ \gamma_{12} \end{vmatrix} = \begin{vmatrix} S_{11} & S_{12} & S_{13} & 0 & 0 & 0 \\ S_{12} & S_{11} & S_{13} & 0 & 0 & 0 \\ S_{13} & S_{13} & S_{33} & 0 & 0 & 0 \\ 0 & 0 & 0 & S_{44} & 0 & 0 \\ 0 & 0 & 0 & 0 & S_{44} & 0 \\ 0 & 0 & 0 & 0 & 0 & S_{66} \end{vmatrix} \cdot \begin{vmatrix} \sigma_1 \\ \sigma_2 \\ \sigma_3 \\ \tau_{23} \\ \tau_{31} \\ \tau_{12} \end{vmatrix} \tag{1.48}
$$

where $S_{66} = 2(S_{11} - S_{12})$

In practical tests on materials, it is usual to measure the tensile strain that results from a uniaxial tensile stress, so that a Young's modulus is obtained. If this is done in the three principal directions in an orthotropic material, three moduli would be obtained; E_1, E_2, and E_3. In addition, the strains transverse to each stress should be measured, to obtain Poisson's ratios, v_{ij}, where

$$
v_{ij} = -\varepsilon_j / \varepsilon_i \tag{1.49}
$$

($i, j = 1, 2, 3, i \neq j$; the material is stressed in the i direction).

Only three Poisson's ratios are required, since

$$
v_{ij}/E_i = v_{ji}/E_j \tag{1.50}
$$

In addition, in order to determine all the elastic properties of the composite, shear tests should be performed to measure the three shear moduli G_{23}, G_{31}, and G_{12}. We can now express $[S]$ in terms of these "engineering constants":

$$
[S_{ij}] = \begin{vmatrix} \dfrac{1}{E_1} & \dfrac{-v_{12}}{E_1} & \dfrac{-v_{13}}{E_1} & 0 & 0 & 0 \\[2mm] \dfrac{-v_{12}}{E_1} & \dfrac{1}{E_2} & \dfrac{-v_{23}}{E_2} & 0 & 0 & 0 \\[2mm] \dfrac{-v_{13}}{E_1} & \dfrac{-v_{23}}{E_2} & \dfrac{1}{E_3} & 0 & 0 & 0 \\[2mm] 0 & 0 & 0 & \dfrac{1}{G_{23}} & 0 & 0 \\[2mm] 0 & 0 & 0 & 0 & \dfrac{1}{G_{31}} & 0 \\[2mm] 0 & 0 & 0 & 0 & 0 & \dfrac{1}{G_{12}} \end{vmatrix} \tag{1.51}
$$

Relatively few elasticity problems have been solved for orthotropic materials, on account of the complexity of the stress–strain relations.

Further Reading

TIMOSHENKO, S. (1958) *Strength of Materials*, Vols. 1 and 2 (D. Van Nostrand, New York).
VAN VLACK, L. H. (1964) *Elements of Materials Science* (Addison Wesley, Reading, Mass).
GORDON, J. E. (1976) *The New Science of Strong Materials* (Penguin Books Ltd., Harmondsworth).

2

Physical Factors Influencing Mechanical Properties

COMPOSITE materials make use of our knowledge of the influence of bond strength on materials strength, which is exerted through the internal structure of the material. Thus it is necessary first to understand the causes of the strengths and weaknesses of traditional materials.

It was only relatively recently realized that there is some connection between chemical bond strengths and the tensile strengths of materials. This is because the relationship is extremely indirect, and a given element or compound can have a great many different strengths according to how a sample of it is made, and the precise nature and distribution of the impurities within it.

A good example of this is iron. A small amount of carbon can greatly influence its strength, even when the carbon has no effect on the nature of predominant interatomic bonds. The effect of the carbon is strongly influenced by the heat treatment the iron (or steel) has undergone, and the strength of the material is also very very sensitive to the prior mechanical treatment, again without any need to change the predominant bonding.

In this chapter, a rough estimate will be made of how strong a material could be, if perfect. This will be followed by a description of the two major sources of weakness in real materials.

2.1. Strength of Solids Calculated from Bond Strengths

A large number of attempts have been made to estimate the strength of materials from the strength of the chemical bonds within them. At first the results obtained seemed so high compared with the known strengths of materials, and varied in such an apparently random way from one material to another, that people were inclined to believe that the strength of solids had little to do with bond strengths.

One of the simplest ways of obtaining a rough estimate of the theoretical strength of a solid is to assume that the variation of interatomic forces with distance can be approximated by a sine curve. Figure 2.1 shows a sine curve which crosses the axis at a_1 and at $a_1 + a_2$. Superimposed on this is a dashed line that indicates a typical force-distance relationship between two atoms, where this deviates significantly from the sine curve. At $x = a_1$ there is no force, and this is the equilibrium position for the two atoms. Compression (negative force) is required to push the atoms closer together than the equilibrium value, and tensile forces are required to separate them.

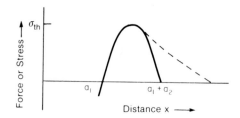

FIG. 2.1. Approximate force–distance relation for two atoms.

If, instead of having two atoms, we have a solid crystal, consisting of vast numbers of atoms (a cubic mm of a solid contains $\sim 10^{20}$ atoms) arranged in some orderly way in rows and planes, we can assume that the force between two planes also follows the sine curve. If we apply a small stress (force per unit area), the distance between the planes will increase, and the rate of increase for increasing force is given by the modulus of elasticity, E (Young's modulus).

A plot of stress, σ, against distance will look the same as the force plot, and will have the equation.

$$\sigma = k_1 \sin \left(\pi (x - a_1)/a_2 \right) \tag{2.1}$$

where k_1 is a constant which can be determined from the slope of the curve at $x = a_1$, where the stress is zero. At $x = a_1 + dx$, the stress is $d\sigma$ and the strain is dx/a_1. We can equate the modulus with the stress/strain ratio. Thus

$$E = a_1 \frac{d\sigma}{dx} \tag{2.2}$$

at $x = a_1$. From equation (2.1)

$$\frac{d\sigma}{dx} = \frac{k_1 \pi}{a_2} \cos \left(\pi (x - a_1)/a_2 \right)$$

so that

$$\frac{d\sigma}{dx} = k_1 \pi / a_2$$

at $x = a_1$. Thus, substituting this into equation (2.2), we find that

$$k_1 = E a_2 / \pi a_1$$

If we now assume that $a_2 \simeq a_1$,

$$k_1 \simeq E/\pi. \tag{2.3}$$

The breaking-strength of the material should be determined by the maximum stress. This is when $x = a_1 + a_2/2$, and has the value $\sigma_{th} = k_1$ (from equation (2.1)). Thus the theoretical strength, using equation (2.3) is

$$\sigma_{th} = E/\pi. \tag{2.4}$$

If the strength were governed by this expression we would expect aluminium and glass

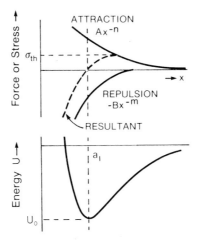

F$_\text{IG}$. 2.2. Force–distance relation for two atoms, and associated energy.

to have roughly the same strength, of about 22 GPa. However, macroscopic pieces of the best aluminium alloys have strengths of only 0.45 GPa, and there seems to be little potential for improving this figure, while the best glass sheets, rods, or tubes have a strength of only about 0.1 GPa.

A better approximation is obtained if we assume that the interactions between atoms are governed by two forces, one attractive and the other repulsive. The force of attraction originates from electrostatic forces (ionic solids), or covalent bonding forces (for elements like carbon or silicon), or metallic bonding forces, or Van der Waals forces. The repulsive force is a measure of the "squashability" of the atoms, and only becomes significant when the atoms are "touching". An expression used to describe these forces is

$$\sigma = Ax^{-n} - Bx^{-m} \tag{2.5}$$

and is shown schematically in Fig. 2.2. Integration of this equation gives an expression for the energy associated with these forces for $U = 0$ when $x \to \infty$:

$$U = -\frac{A}{n-1}x^{-n+1} + \frac{B}{m-1}x^{-m+1}$$

This has a minimum value when $x = a_1$ of

$$U_o = \frac{-A(m-n)}{(n-1)(m-1)}a_1^{-(n-1)}$$

where a_1 is the equilibrium separation of the atoms i.e. where the attraction and repulsion counterbalance each other, so that $\sigma = 0$. To dissociate the material we need to supply this energy U_o. Thus we can obtain information about the constants in equation (2.2) by measuring the heat required to vaporize or sublime the material.

We can use equation 2.5 to estimate the theoretical strength, if we have values for m, n, and the modulus of elasticity. As before, $E = x d\sigma/dx$ at $x = a_1$, and $\sigma = 0$ at this point.

Thus

$$B = Aa_1{}^{m-n} \tag{2.6}$$

and

$$E = a_1(-nAa_1{}^{-n-1} + mBa_1{}^{-m-1})$$

which can be rearranged to give

$$A = Ea_1^n/(m-n). \tag{2.7}$$

We have now evaluated A and B in terms of E and a_1 together with m and n. The maximum strength will be when the stress reaches its maximum value, i.e. when $d\sigma/dx = 0$. This is when,

$$x^{m-n} = mB/nA.$$

Substituting for A and B using equations (2.7) and (2.6), and taking the $(m-n)$th root gives

$$x = a_1\left(\frac{m}{n}\right)^{\frac{1}{m-n}}$$

At this value of x the stress (equation (2.5)) is σ_{th} where

$$\sigma_{th} = Aa_1^{-n}\left(\frac{n}{m}\right)^{\frac{n}{m-n}} - Ba_1^{-m}\left(\frac{n}{m}\right)^{\frac{m}{m-n}}$$

Substituting for B from equation (2.6) and for A from equation (2.7) we obtain

$$\sigma_{th} = \frac{E}{m-n}\left\{\left(\frac{n}{m}\right)^{\frac{n}{m-n}} - \left(\frac{n}{m}\right)^{\frac{m}{m-n}}\right\}.$$

But $m/(m-n) = n/(m-n) + 1$, therefore

$$\sigma_{th} = \frac{E}{m-n}\left(\frac{n}{m}\right)^{\frac{n}{m-n}}\left(1-\frac{n}{m}\right)$$

and so finally,

$$\sigma_{th} = \frac{E}{m}\left(\frac{n}{m}\right)^{\frac{n}{m-n}}. \tag{2.8}$$

Thus, the theoretical strength can be calculated if E, m, and n are known.

For alkali halides $n = 2$ (electrostatic attraction falls off as the square of the distance) and m is about 10. So we would expect an alkali halide to have a theoretical strength of about one-fifteenth of its Young's modulus. The value for potassium chloride is 1.3 GPa, which is vastly greater than the observed strength of a crystal of the material, which can be less than 1 MPa. In the case of metals also, the values of m and n are such that the theoretical strength is about one-fifteenth of the Young's modulus. For iron, with $n = 4$ and $m = 7$ the value is 14 GPa, which should be compared with 4.2 GPa obtained with very strong steel wires.

The theoretical values calculated so far are for cleavage failure of the material. Failure by shear may also be possible and it is worth while examining the stress required for this, to determine whether shear failure could occur at a lower stress than tensile failure.

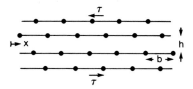

FIG. 2.3. Shear of planes of atoms over one another.

To calculate approximate values for the theoretical shear strength of materials, we will again assume a sine law variation of force with distance. We will consider two adjacent planes in a crystal, separated by a distance h, and containing atoms spaced at distance b (Fig. 2.3). As a shear displacement x occurs, the stress will be given by

$$\tau = k_2 \sin (2\pi x/b). \tag{2.9}$$

(We assume that the stress has a minimum when the atoms are opposite each other, as well as when they are in alternate positions as shown in Fig. 2.3). For small x the displacement will be governed by the shear modulus, G, i.e.

$$G = h\frac{d\tau}{dx}$$

for $x \simeq 0$, but from equation (2.9),

$$\frac{d\tau}{dx} = \frac{2\pi k_2}{b} \cos (2\pi x/b)$$

so that

$$k_2 = Gb/2\pi h.$$

The maximum force occurs when $x = b/4$ and then has the value k_2. Thus the theoretical shear strength τ_{th} is

$$\tau_{th} = Gb/2\pi h. \tag{2.10}$$

For face centred cubic metals, $b = a_o/\sqrt{6}$ and $h = a_o/\sqrt{3}$ where a_o is the length of the unit cell edge. We conclude, therefore, that

$$\tau_{th} \simeq G/9$$

If the material is in tension, this shear stress will be produced by a tensile stress of $2G/9$, and if the material is assumed to be isotropic, this comes to about $E/12$. Thus equation (2.10) gives a similar answer to equation (2.8). To determine whether shear failure or tensile failure is most likely to occur requires a much more detailed analysis. (This may be found in "Strong Solids"; see the list of further reading at the end of the chapter.)

We conclude, finally, that the theoretical strength is some fraction of the modulus. A representative value, useful for comparison with practical strengths, is about $E/15$.

2.2. Dislocations

Although, as indicated in the previous section, the bonding forces between atoms should be able to impart very great strength to materials, the orderly arrangement of the atoms in solids makes deformation and failure possible at very low stresses. Within limits, the more perfect the structure is the weaker it is. This is particularly true with pure metals. The reason for this is that even the most regular arrangements still have occasional imperfections. In a crystal containing 10^{20} to 10^{22} atoms there will always be many places where atoms are missing, and other places where there are too many atoms. There will be still other places where impurity atoms distort the structure, even in the purest of materials.

These "point defects", however, are not the cause of the surprisingly easy deformation found with many pure metals. The main cause are lines of defects called dislocations.

The simplest type of dislocation to visualize is the edge dislocation, shown in Fig. 2.4. The undistorted, or perfect structure, is shown at the left. The atoms are located at the crossing-points of the lines, and can be visualized as spheres, connected by springs to represent the bonding forces. The edge dislocation is equivalent to the insertion of an extra half-plane into the structure. This causes considerable distortion close to where the plane ends (WX) but relatively little elsewhere.

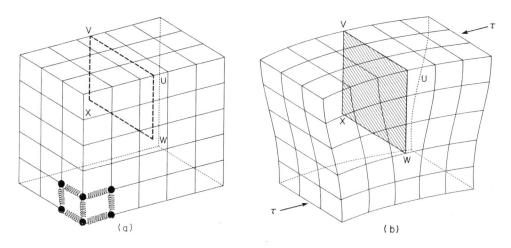

FIG. 2.4. Edge dislocation: (*a*) position of extra plane, (*b*) distortion resulting from the presence of the extra plane.

This extra half plane can move very easily across the crystal when there is a shear stress, τ, and results in the deformation shown in Fig. 2.5.

The inverted T symbolizes the dislocation. The horizontal part indicates the plane on which slip takes place (the slip plane), and the vertical part indicates the relative position of the extra plane of atoms. Figure 2.6 shows the deformation produced by the motion of many dislocations.

The calculation of the stress required for dislocation movement, the Pieirls stress, is rather difficult. The answer comes to zero for a straight dislocation unless terms of very small magnitude are included in the calculations (all the larger terms cancel out). Its value is very sensitive to the force–distance relation between neighbouring atoms. Figure 2.7 is a schematic drawing of the atomic positions around an ideal edge dislocation in a simple cubic lattice. If the dislocation width, W, is defined as the total distance in the slip direction,

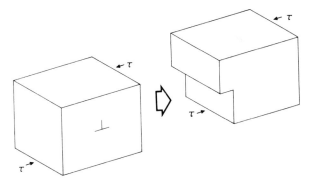

FIG. 2.5. Deformation due to movement of an edge dislocation to the surface.

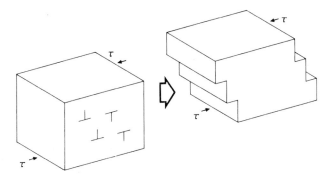

FIG. 2.6. Deformation due to the movement of many edge dislocations to the surface.

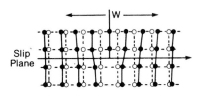

FIG. 2.7. Atomic positions around an edge dislocation.

along the slip plane, over which the displacement of atoms is greater than one-half of the maximum displacement, calculations suggest that the Peierls stress depends very strongly upon W.

For example, it has been suggested that

$$\tau = \frac{E}{1 - \nu^2} e^{-2\pi W/b} \tag{2.11}$$

where b is the displacement associated with the dislocation, called the Burgers Vector.

(Being a vector, it represents the direction as well as the magnitude of the dislocation.) It is usually the distance between neighbouring planes in the crystal lattice, or some fraction thereof. Equation (2.11) gives a very low result for metals. They have W roughly equal to three times b, so that $\tau \sim E/10^6$.

Unfortunately, although dislocations can be seen in the electron microscope, the exact positions of atoms around them cannot be determined, because even in images which appear to show atomatic positions, the picture is a smeared-out one, with atoms in many planes (1000 or more) contributing to it.

Calculations of W have not been any more successful than direct observations, and since the Peierls stress is inversely proportional to an exponential function of W, even the order of magnitude of the Peierls stress is subject to considerable uncertainty. Experiments with very pure metals have shown that it can be very low; the more sensitive the apparatus, the lower the result obtained for the stress at which flow starts.

After a small amount of deformation has occurred the flow stress starts to increase. This is due to the generation of large numbers of new dislocations. They interfere with each other, making movement progressively more and more difficult. This process occurs with most metals and is called work hardening.

Dislocations may be regarded as units, or quanta, of slip. In real materials the displacements suffered by the atoms close to dislocations are much more complicated than the simple picture given here. Two limiting types of dislocation have been identified, the edge dislocation, which is the one already discussed, and the screw dislocation shown in Fig. 2.8. With the edge dislocation the slip or displacement is perpendicular to the dislocation line. With the screw it is parallel to the dislocation line. Normally dislocations are a mixture of edge and screw, and to make matters even more complicated, they often sub-divide themselves into partial dislocations. These involve smaller displacements of the atoms. In addition, the simple cubic structure pictured here does not adequately represent the arrangement of atoms in crystal lattices. The simplest crystals can be represented by cubic units, but with additional atoms at the centres of each face of the cube. The edge dislocation is thus not as simple as indicated in the drawings, and its movement is more tortuous than the linear movement indicated.

The complicated nature of dislocations, their generation and their interactions, has made quantitative assessments of mechanical properties very difficult. But they have been

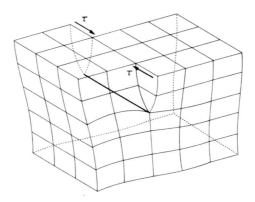

FIG. 2.8. Screw dislocation.

very useful in giving insight into the mechanical behaviour of materials, especially metals. They explain, in a qualitative fashion, why metals deform so easily; dislocations can be seen to move in the electron microscope under forces which are very small compared with atomic binding forces. The difficulty dislocations have in crossing crystal boundaries into neighbouring crystals explains why fine-grained materials usually have higher yield stresses than materials with large crystals. Their interactions with precipitates within metals helps to account for the improvements in mechanical properties than can be obtained by using alloys rather than pure metals, and suggests ways in which still better alloys may be developed.

Softening as well as hardening effects can be explained. Annealing causes the disappearance of large numbers of dislocations, as well as some recrystallization. Thus, when a work-hardened metal is annealed and subsequently deformed, the few dislocations remaining can move with relatively little interference from other dislocations.

These are just a few examples of the long list of effects which can be understood, or at least in which insight has been gained, by the use of explanations involving dislocations.

2.3. Notches and Cracks

To obtain strong metals it is necessary to interfere with dislocation motion so that plastic flow does not occur at low stresses. Our understanding of how to do this has improved steadily in recent years, but it has been found that there is a limit to this method of improving materials. Aluminium alloys with yield strengths greater than 0.45 GPa and steels with yield strengths greater than 1.4 GPa have been known for 50 years. But they may only be used in a few carefully chosen applications and then only with great care. The reason for this is that when the ratio of the yield stress to the modulus reaches a certain value, the material no longer has sufficient ductility to make the necessary internal adjustments to alleviate the effect of the excess stresses that are to be found around the holes, notches, and cracks that are always present in most structures, even the most carefully built ones.

2.3.1. Stress Concentrations

Inglis originally showed in 1913 that the stresses close to the tips of cracks could be much higher than elsewhere in a piece of material. He did this by determining the stresses around elliptical holes. The ellipse is a particularly suitable shape for this purpose, because at one extreme value of the ellipticity (the ratio of major axis to minor axis) it is a circle, and at the other extreme it is an infinitely thin crack. Thus it can be used for cracks ranging from sharp ones of the utmost severity to mild cracks and round holes. In addition, the stresses and strains can be expressed completely in algebraic form (though they are rather cumbersome).

Figure 2.9 represents an elliptic crack in a material under a stress σ_x normal to the crack plane.

Figure 2.10 shows how the stresses are concentrated around the crack tip. The lines are loci of constant stress concentration, and the numbers indicate the amount by which the stress is multiplied. For such a sharp crack as this, where the ellipticity is 100 (i.e. the major axis of the ellipse is 100 times longer than the minor axis) the curvature at the crack tip is

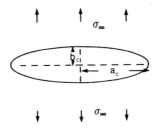

FIG. 2.9. Elliptic crack.

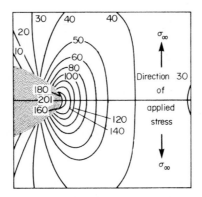

FIG. 2.10. Stresses near the tip of an elliptic crack. (After Gordon, J. E. (1964) *Proc. Roy. Soc.* **A282**, 508.)

very sharp, and the stresses are correspondingly very high. At the crack tip the stress is 201 times greater than the applied stress.

In addition to the stresses in the same direction as the applied stress, there are stresses in directions at right angles to this. Figure 2.11 shows the stresses in the direction of the major axis of the ellipse.

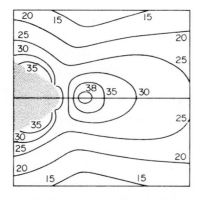

FIG. 2.11. Stresses at right angles to the stress applied to an elliptic crack. (After Gordon, J. E. (1964) *Proc. Roy. Soc.* **A282**, 508.)

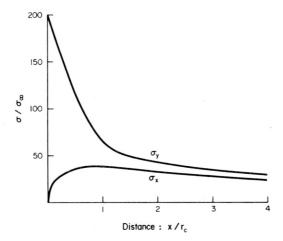

FIG. 2.12. Stresses near a crack tip. The applied stress (σ_x) is in the same direction as σ_y; a_c/b_c = 100. r_c is the radius of the crack tip.

The stresses in the crack plane are also shown in Figure 2.12. Here σ_y is the stress in the direction of the applied stress (σ_x) and σ_x is the stress in the direction of the major axis of the ellipse. There are also stresses normal to the plane of the sheet.

Although the stresses at right angles to the applied stress are much smaller than those in the same direction as the applied stress, their effect is very important. The material close to the crack tip is subject to tensile forces in all directions, and this triaxial stress system inhibits shear flow. When the sheet is thin this is not very serious, the stress normal to the sheet being relatively small. Thus, if the material is ductile enough, some plastic flow can take place which will relieve the stresses and round off the crack tip. However, if the sheet is thick, and not particularly ductile, high stresses are developed normal to the sheet, stress relaxation is not so likely to take place, and brittle fracture is facilitated.

The maximum stress at the crack tip in a sheet, under tensile stress σ_x, along the direction of the minor axis of the ellipse (the stress being applied a long way away from the crack), is given by the equation

$$\sigma_{tip} = \sigma_x (1 + 2e) \tag{2.12}$$

where e is the ellipticity, a_c/b_c (see Figure 2.9). This formula only gives accurate values when the width of the sheet is very much greater than a_c, and the crack is not close to the edge of the sheet. If the radius of the crack tip is r_c, then the expression can also be written

$$\sigma_{tip} = \sigma_x (1 + 2 \sqrt{(a_c/r_c)}) \tag{2.13}$$

This equation has been found to give about the correct answer for the stresses at the tips of cracks in the surfaces of materials if the 1 is neglected. It is not even necessary for the cracks to be elliptical. Figure 2.13 shows examples of cracks and steps to which this formula has been found to apply with reasonable accuracy.

For very sharp cracks, or surface steps, very high stress concentrations are possible. If the crack or step has atomic dimensions at the tip, i.e. $r_c \simeq 0.3$ nm, the height of the step or depth of the crack only needs to be 0.75 μm for a stress concentration factor of 100 to exist at the crack tip. Thus a material in which dislocations cannot move, or which contains no

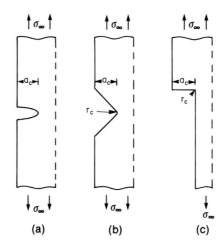

FIG. 2.13. Cracks and steps with stresses that obey the equation $\sigma_{tip} = \sigma_{\infty}(1 + 2e)$.

dislocations and so cannot yield under stress, is in danger of having the atomic bonds at the crack tip stressed to the breaking-point when relatively small forces are applied to it.

This appears to be what happens in the case of glass. Glass does not have a regular structure, so dislocations cannot exist within it. It is very hard to produce any plastic effects in glass at room temperature; even a very sharp diamond, after being pressed into glass, leaves only a very tiny impression. The surface of ordinary glass is found to have millions of tiny cracks in it. These cracks form spontaneously, probably due to water vapour in the air, but are also caused by the surface coming into contact with hard materials during manufacture. Since glass has a theoretical strength of about $E/10$ or 7 GPa, and yet its strength is normally only about 70 MPa, we might conclude that the cracks need only have a length of about a micron if the tip has atomic dimensions. However, we also need to consider whether crack propagation is energetically favourable, since if no energy is released by the fracture process, it cannot occur spontaneously.

2.3.2. Energy Considerations

In 1920 A. A. Griffith explained why the tendency of a material to fracture does not depend only on the severity of the crack. There is a size effect also; thus for two notches having the same value of a_c/r_c but different sizes, the larger notch is much more likely to cause failure than the smaller one.

Griffith went on to show that energy must be continuously released for crack propagation to occur. Where this energy comes from appears to be a much misunderstood matter.

A cracked sheet, loaded by, for example, a hanging weight, has more elastic strain energy than a similar sheet without the crack. (The elastic strain energy of an element of material is one-half of the product of the stress, the strain, and the volume of the element.) The extra strain energy results from the stress concentration, which raises the average stress level, and also increases the strains. This extra energy is provided by the loading system, the hanging weight in our example. We find that the cracked sheet is more easily deformed (more compliant) than the uncracked one. Thus the weight moves further down when loading it, providing the extra energy needed, with an equal amount left over.

It was Griffith who showed that the potential energy of the system applying the forces (the weight) must decrease by exactly twice the extra energy due to the stress-concentrating effect of the track. Thus the overall decrease in energy, i.e. the decrease in potential energy minus the increase in strain energy, is exactly equal to the increase in strain energy of the body due to the presence of the crack.

When other loading methods, such as fixed grip systems are used, the result is the same, but the analysis is slightly more complicated.

Some extra energy is required for the production of a crack, however. This is the work done in separating the atoms on one side of the crack face from those on the other side. In an ideal non-yielding solid, this work is equal to the surface energy of the solid, ψ, multiplied by the total area of surface produced. The total extra energy due to a crack with vanishingly small width, b_c, in a thick sheet of the ideal solid is

$$U = 4a_c\psi - \frac{\pi\sigma_\infty^2 (1 - v^2)a_c^2}{E} \qquad (2.14)$$

where v is Poissons ratio, and E the modulus of the solid. The first term is the energy absorbed in the production of new surface, and the second term is the energy released due to the loading system. (The second term is the same for a constant weight loading, and for fixed grip loading.)

For a crack to extend, the rate of release of elastic energy as a_c increases must exceed the rate of increase of the total surface energy for the same increase in a_c. The limiting case is when the two are equal. This is when $dU/da_c = 0$. Since

$$\frac{dU}{da_c} = 4\psi - \frac{2\pi\sigma_\infty^2 (1 - v^2)a_c}{E}$$

this is when

$$\sigma_\infty^2 = \frac{2E\psi}{\pi(1 - v^2)a_c}. \qquad (2.15)$$

Equation (2.15) predicts that the failure stress should be inversely proportional to the square root of the crack length. Griffith showed that this relationship was obeyed by glass, and that an equation similar to (2.15) was quantitatively correct if ψ was equated with surface energy. We now know that there were some slight errors in his analysis, and that ψ calculated from fracture experiments with glass is three to ten times higher than the surface energy.

We can now calculate the crack length required for glass to fracture at its normal breaking-strength, i.e. 70 MPa. We use the value of E from Table 1.1 and half the work of fracture (Table 1.2) for ψ (ψ is only half the work to fracture 1 m^2 of material, since the fracture produces 2 m^2 of new surface). Equation (2.15) produces the result $a_c = 50\ \mu m$.

At the end of Section 2.3.1 we showed that if cracks in glass had atomically sharp tips, a crack only one micron long in a sheet stressed to 70 MPa would generate a stress equal to the theoretical strength at the tip. Clearly, if the crack has to be 50 microns long for failure to be energetically favourable, the crack tip radius must be several interatomic distances in order that the theoretical strength is not exceeded.

Since a crack in such a brittle material as glass is not atomically sharp, we need to examine the concept of surface energy in more detail and see if it can indeed be used for ψ in the Griffith equation (equation (2.15)).

2.3.3. Surface Energy

The Griffith theory described above requires that the elastic energy released exceeds the surface energy required to produce the new fracture faces. The surface energy for solids, though larger than for liquids, is rather small, as can be shown by calculating it directly from the interatomic forces, as follows.

The surface energy is the energy associated with the unsatisfied interatomic bonds at the surface. These can be: (a) primary bonds, for example the electrostatic forces at the surface of a freshly cleaved sodium chloride crystal in a perfect vacuum; or (b) secondary bonds such as the hydrogen bonds that hold water molecules together, and create the surface tension that keeps the surface area of water droplets as small as possible.

We may calculate the surface energy of a solid by estimating the work needed to separate the atoms on either side of a plane in the solid. They must move sufficiently far apart that the interatomic forces have fallen to zero. The work required to do this is equal to the area under the force–distance curve in Fig. 2.2, between the equilibrium distance, a_1, and infinity. For simplicity, we will consider unit area of cross-section. This will produce two surfaces. Thus

$$2\psi = \int_{a_1}^{\infty} \sigma \, dx \tag{2.16}$$

substituting for σ using equation (2.5) and dividing both sides by 2 we obtain

$$\psi = \frac{1}{2} \int_{a_1}^{\infty} (Ax^{-n} - Bx^{-m}) \, dx$$

Integrating and evaluating this expression gives

$$\psi = \frac{1}{2}\left(\frac{Aa_1^{1-n}}{n-1} - \frac{Ba_1^{1-m}}{m-1} \right)$$

and substituting $B = Aa_1^{m-n}$ and $E = A(m-n)a_1^{-n}$ we find that

$$\psi = \frac{Ea_1}{2(m-1)(n-1)} \tag{2.17}$$

We can now estimate the values of ψ_s for materials for which m and n are known, together with E and a_1.

For many materials the interatomic separation is roughly 0.25 nm and $(m-1)(n-1)$ is of the order of 20. For metals and inorganic crystals E is in the range 10 to 1000 GPa. Thus ψ should be in the range 0.05 to 5 J m^{-2} for metals and inorganic materials, and somewhat less for polymers due to their lower moduli. Iron, for example, with $m = 7$ and $n = 4$ should have a surface energy of about 1.5 J m^{-2}.

These values are very low compared with most of the works of fracture given in Table 1.2. Maraging steel, which is mostly iron, has a result which is 40,000 times higher.

Although 2ψ rather than ψ should be compared with the work of fracture, this still leaves a factor of 20,000 to account for.

2.3.4. Fracture Toughness

About 10 years after Griffith published his work on glass G. R. Irwin, observing that metals obeyed the same expression for fracture failure; i.e. $\sigma_\infty^2 a_c = $ constant, used $\mathbf{G}$, the work of fracture, instead of 2ψ. Thus,

$$\sigma_\infty^2 = \frac{E\mathbf{G}}{\pi(1 - v^2)a_c} \tag{2.18}$$

It is clear that $\mathbf{G}$ is analogous to the surface energy. Since it is usually so much greater than 2ψ we conclude that fracture failure involves a great deal more than simply separating the appropriate atoms. That this is so can easily be demonstrated by examining the fracture surfaces of any fairly tough piece of material. It will be observed that a great deal of plastic work has been done on the metal near the new surface.

The fracture toughness for failure of the type we have described is defined by the expression:

$$K_{1c} = \sigma_\infty \sqrt{(\pi a_c)} \tag{2.19}$$

when the applied stress, σ_∞, is sufficiently great to break a piece of the material, having a crack of length $2\hat{a}_c$ in the centre, or of length a_c if it is at the edge. When σ_∞ is smaller than this,

$$\mathbf{K}_1 = \sigma_\infty \sqrt{(\pi a_c)}$$

and $\mathbf{K}_1$ is the stress intensity factor. $\mathbf{K}_{1c}$ is often referred to as the critical stress intensity factor. (There are also a $\mathbf{K}_{2c}$ and a $\mathbf{K}_{3c}$ for cracks loaded in shear instead of in shear instead of in tension). $\mathbf{K}_{1c}$ and $\mathbf{G}$ are thus related:

$$\mathbf{G} = (1 - v^2)\mathbf{K}_{1c}^2/E \tag{2.20}$$

The fracture toughness is measured by determining the compliance of a specimen at the instant a crack of known size in it starts to propagate. Alternatively $\mathbf{G}$ may be measured by determining the work required to propagate a crack slowly through a known distance in the material. The stress may be applied by tension, or flexure of the specimen. It is essential that the crack is extremely sharp, otherwise $\mathbf{G}$ and $\mathbf{K}_{1c}$ are overestimated. This is ensured, in the case of metals, by notching the metal, and then fatiguing it so that the notch extends a few mm further. Corrections are made for notches that cut through a significant fraction of the cross section of the specimen, and precautions have to be taken to ensure that the specimen is sufficiently thick for the test to be valid. (The tougher the material, the thicker the specimen has to be.)

Older methods of measuring toughness are still widely used, especially for polymers and composites. Two important methods involve impact of a heavy hammer against a notched bar. The Charpy method uses a pendulum to strike a specimen in such a way that it is loaded in three-point bending. The energy lost by the pendulum is measured. For polymers and reinforced plastics the result is usually given as so many foot pounds of energy per inch width of notch. The specimen has a standard thickness in the direction in which it is struck (12.7 mm), and a standard notch depth of 2.54 mm. For 1-inch width of notch the broken surface has an area of 0.258×10^{-3} m^2 (0.4 sq. inches), so 1 ft-lb/inch of notch is equivalent to 5.26k Jm^{-2}. For metals the specimen has a standard size of 10 mm $\times$ 10 mm and the notch depth is 2 mm. Sub-size specimens (each dimension halved) are also sometimes used.

The Izod test operates on the same principle, except that the specimen has one end fixed in a clamp, and the pendulum breaks off the protruding end. The specimen dimensions are the same as for the Charpy test.

(Both tests are sometimes carried out on "unnotched" specimens. This is usually done with brittle materials, since otherwise the energy absorbed is too small to be measured accurately. With reinforced plastics this is done simply by reversing the notched specimen, so that the notch tip is under compression; with ceramics a round unnotched bar is used.)

Both tests can be improved by providing sensors to measure the forces exerted by the impacting hammer. These are called instrumented impact tests. A plot of force during impact vs. time is obtained using a fast recorder such as an oscilloscope. Use of these plots can eliminate some of the inaccuracies arising from the method, and $\mathbf{K}_{1c}$ values may be obtained directly.

Further improvement has resulted from making measurements at different notch depths and plotting the $\mathbf{K}_{1c}$ values obtained as a function of a_c. ($\mathbf{K}_{1c}$, of course, has to be corrected for each a_c value, since a_c is a significant fraction of the total cross section in these tests). The plot normally has a plateau region where $\mathbf{K}_{1c}$ is independent of a_c. The value of $\mathbf{K}_{1c}$ in this region is the value required.

Another type of test is the Slow Bend Test. In this test the specimen is notched in such a way that the crack propagates slowly across the remainder of the cross-section. The work of fracture is calculated from the area under the force—distance curve. If unstable fracture occurs and the crack propagates quickly across the specimen the test is invalid, and a different shape of notch must be used.

2.4. Practical Limits of Strength

We have shown that the theoretical strength is very large and determined by the chemical bond strength. However, such high strengths are not normally achieved in practice for one of two reasons. Either dislocations can move at low stress, leading to early failure by slip and plastic deformation, or the material is brittle, and can fail at very low stresses by the propagation of cracks which start at tiny imperfections, often too small to be seen except with powerful microscopes.

With metals it is possible to interfere with dislocation movement, and so increase the strength. However, this makes the material more susceptible to brittle failure. This is illustrated in Fig. 2.14 which is a plot of yield stress vs. Young's modulus. A number of values have been plotted for metals to indicate the range of yield stresses that can be achieved. The notched area indicates the transition region between ductile and brittle materials. Materials such as glass, quartz, and diamond are well in the brittle region. (They can be ductile at high temperature, however, since this decreases the yield stress.)

It is clear from Fig. 2.13 that if we want high strength coupled with adequate toughness we must use a material with a high modulus. The practical maximum strength of steels is

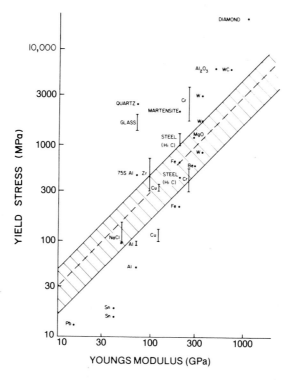

FIG. 2.14. Yield stress vs E indicating ductile-brittle transition. (After Piggott, M. R. (1975) *Int. J. Fracture* **11**, 479.)

about three times that of aluminium, and that of tungsten is higher still. In fact, a dimensional analysis predicts that serious brittleness occurs when the yield strain exceeds a constant value. Experience shows that it should not exceed 0.007.

Also, we cannot allow the strains to become too high, as is graphically demonstrated in Fig. 1.7.

Usually we want a material that is light as well as strong and tough. Unfortunately, this does not appear to be possible with traditional load-bearing materials, since E/ρ is approximately constant for them (see Table 1.3).

A new approach is clearly needed. Fibre reinforced materials provide this.

Further Reading

KELLY, A. (1973) *Strong Solids* (Oxford University Press, London).
HULL, D. (1968) *Introduction to Dislocations* (Pergamon Press, Oxford).
ROLFE, S. T. and BARSON, J. M. (1977) *Fracture and Fatigue Control in Structures* (Prentice-Hall, New Jersey).
Fracture Toughness Testing and its Applications, ASTM STP 381 (1965) (ASTM, Philadelphia).

3

Fibres, Whiskers, and Platelets

WE HAVE shown in Chapter 2 that there is a limit to how far we can go with traditional materials in the search for high strength with adequate toughness and low density. In this chapter we will show that slender forms of material provide an alternative of great potential.

3.1. Slender Forms of Material

If we make a piece of material small enough, we can suppress the weakening due to fracture, and that due to disolution movement.

Consider fracture first. To achieve a strength of about the theoretical value, i.e. $E/15$, we must make the cross section of the material such that no diameter can contain a crack of length longer than a_c where

$$\left(\frac{E}{15}\right)^2 = \frac{EG}{\pi a_c(1-v^2)}$$

(This is equation (2.18) with $E/15$ replacing σ_∞.(G is the work of fracture). For $v \simeq 0.3$,

$$a_c \simeq 80G/E \tag{3.1}$$

For alumina this comes to about 4 nm using the data given in Chapter 1. Thus an alumina fibre of 4 nm diameter should have the theoretical strength.

Now consider failure resulting from dislocation movement. Ductile failure does not result directly from the movement of dislocations initially present. Instead, a dislocation generator or multiplier is needed. The Frank-Read source is the most important of these, and consists of a dislocation which is held at its ends. (It can be held by other dislocations oblique to the slip plane, or by precipitates, etc.) Figure 3.1 shows a dislocation fixed at A

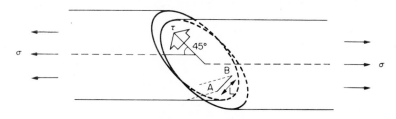

FIG. 3.1. Frank–Read source in a rod. Dislocation is fixed at A and B.

and B. The stress required to generate dislocations depends on the length AB ($= L$), and is given by,

$$\tau = 2Gb/L \tag{3.2}$$

b being the Burgers vector and G the shear modulus.

Suppose this dislocation is in a slip plane at $\pi/4$ to the applied stress, as shown in Figure 3.1. It will then permit slip at the lowest possible stress, and we can write

$$\sigma = 2\tau = 4Gb/L$$

Thus for $\sigma = E/15$, the theoretical strength, $L \sim 24b$ for materials with $G \simeq E/2.5$. Now $b \sim 0.25$ nm, so a ductile fibre with a diameter of about 6 nm should have the theoretical strength.

We, therefore, conclude that slender forms of material can achieve the theoretical strength. The dimensions required are somewhat impractical, though.

Fortunately, when we examine fine forms of material we find that they can be extremely strong. Fibres of glass with very smooth surfaces can be made quite easily. Although flaws are normally present, the strength of these fibres is very great: with pure silica it can reach 6 GPa at 20 C and nearly 10 GPa at -196 C, when free of flaws, while normal production glass fibres have strengths of 3 GPa or more. When fine wires are made they are intensely worked. This causes a great deal of disorganization of the crystal structure, so that dislocations cannot multiply or move, except at very high stresses. At the same time the surface of the wire is quite smooth, so that fracture failure also is inhibited. Thus fine wires can also be made which are very strong indeed.

Potentially the most promising material for high strength, coupled with low density and high modulus, is the single crystal whisker. These are about 1 nm in diameter and have lengths of up to a few mm. Their faces are crystallographic planes, so that they have rectangular or diamond-shaped cross-sections. The surfaces are generally very smooth so that fracture is suppressed, and they usually contain just one screw dislocation, which is near the centre, and parallel to the fibre axis. Their strength can be quite close to the theoretical strength, and their stress–strain trajectories show slight curvature near the breaking-stress, even while they are still perfectly elastic. This is because the interatomic force–distance curve, Fig. 2.2, is only linear at low strains.

Still another form of strong material is the microplate, or platelet. Quite a number of compounds can be made to grow in laminar form, and mica is a well-known example. Figure 3.2 shows another example of this type of material. Although these compounds are potentially strong because dislocations are not usually mobile in them of 20 C, they must have smooth edges to have high strengths.

The importance of size and shape is illustrated in Fig. 3.3 which shows the strength of SiC in various forms.

3.1.1. Choice of Materials

Although slender forms of material can be very strong, to use them we shall need to put them into some type of matrix which will transfer the loads to them, and should ideally not allow them to transmit their inherent brittleness to the composite structure. If we assume for the moment that this can be done, we are free to choose our materials so that we can produce slender forms with maximum strength and modulus and minimum density.

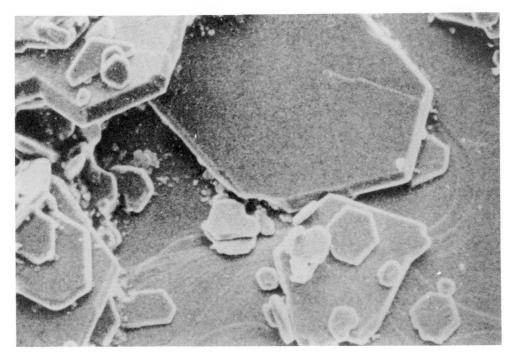

FIG. 3.2. Chromium oxide platelets. (Courtesy of Woodhams, R. T., University of Toronto.)

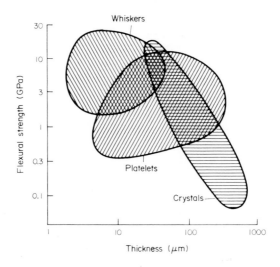

FIG. 3.3. Strength of various forms of silicon carbide.

For light weight we choose elements in the first two rows of the periodic table. We can use them either in elemental form, or as compounds. The elements or compounds should preferably be covalently bonded for the highest bond strengths in two or three dimensions.

Elemental carbon is one of the best examples. The three-dimensional covalently bonded structure of diamond has a very high modulus and a very low density. So does the two-dimensional covalently bonded structure of graphite, though this is stiff and strong only in one plane. Some electrovalent compounds are also potentially very strong. Thus, as candidate materials we have B, C, $B_4 C$, $B_{13}O_2$, B_6Si, AlB_3, SiC, BeO, and Al_2O_3. These all have moduli in the range 400–1000 GPa, and low densities. They also have very high melting- (or sublimation) points.

While these elements and compounds are best in theory, in practice many other materials can be made in slender form, having excellent properties. An important consideration is the cost of making the fibres, and making high quality composites from them. In fact, a very wide range of fibres are available.

3.2. Polymer Fibres and Metal Wires

Polymer fibres and metal wires have been used by man for thousands of years. Many natural fibres have excellent properties, especially when their low densities are taken into account. Table 3.1 lists the properties of some polymers and metals. It can be seen that the best natural polymers have strengths which are as much as one-fifth of the theoretical strength (assuming this is $E/15$).

The chief drawbacks of natural fibres are that they are affected by water, which reduces their strength and stiffness, and are subject to attack by fungi.

Synthetic polymers were, until recently, characterized by very low moduli. Whereas flax and wood cellulose fibres have Young's moduli near 100 GPa, nylon and terylene have moduli between 1 and 3 GPa (see Table 3.1). Recent attempts to produce synthetic fibres with high modulus have met with some success, and aromatic polyamides, under the registered name of Kevlar, have been made with moduli of up to 130 Gpa. The stress–strain curves for most synthetic and natural polymers are far from linear. Figure 3.4 gives some examples.

The stress–strain curves for the Kevlars are, by contrast, almost linear up to the breaking-point. There are three types of Kevlar marketed. One variety, just called Kevlar, is designed for use in tyres and can be used in place of steel in radial tyres. A very similar fibre, Kevlar 29, is used for ropes. For fibre reinforcement of polymers Kevlar 49 is used. This has a modulus of about twice that of Kevlar 29, and about the same strength. It is very tough, which puts it in a class by itself among high-performance fibres, but it has poor compressive properties. It is normally made with a diameter of 12 microns. The Kevlar referred to subsequently in this text is Kevlar 49.

Both natural and synthetic polymers are very sensitive to heat. Few can withstand temperatures much in excess of 150 C, and marked loss in strength occurs with some of them (especially nylon) below 100 C. Kevlar is completely stable at 150 C, but loses strength slowly at 200 C, while at 250 C, after 8 hours, its strength has decreased to about 70% of its initial value, and it continues to decline at a decreasing rate thereafter. It carbonizes at about 425 C. At liquid nitrogen temperature it retains full strength and toughness. It has a small negative axial coefficient of thermal expansion ($-2.0\,\mu K^{-1}$).

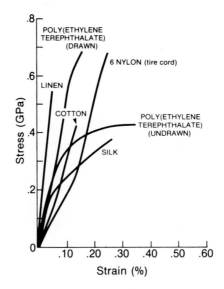

FIG. 3.4. Stress–strain curves for polymer fibres. (After Billmeyer, F. W. (1971) *Textbook of Polymer Science*, Wiley-Interscience, New York.)

Most polymers, however, have positive, and rather large, coefficients of thermal expansion in the range 30–100 μK^{-1} (see Table 5.6), and Kevlar's radial expansion is 59 μK^{-1}.

Although some polymer fibres are quite cheap, the stronger and more heat-resistant ones are quite expensive. Nylon, for example, costs about $1/kg while Kevlar costs about $18/kg.

Metal wires have higher strengths and moduli than most polymers. Table 3.1 gives a few examples. Their densities are, of course, also much greater than those of polymers, so that the specific strengths (strength/density) are comparable. Steel wires are much weakened by

TABLE 3.1.

	Material	Density (Mg m^{-3})	Strength (GPa)	Modulus (GPa)
	Cotton	1.50	0.35	1.1
	Flax		0.9	110
Natural polymers	Silk	1.25	0.5	1.3
	Wool	1.3	0.36	6
	Wood (Kraft paper)	~ 1.0	0.9	72
	Cellulose (Fortisan)	1.52	1.1	2.4
Synthetic polymers	Polyester (Terylene)	1.38	0.6	1.2
	Nylon	1.14	0.8	2.9
	Kevlar 49	1.45	3.6	130
	Beryllium	1.8	1.3	315
	Molybdenum	10.3	2.1	343
	Nickel alloy (Rene 41)	8.2	2.3	220
Metals	Steel	7.9	4.2	210
	Tungsten	19.3	3.9	411
	Titanium alloy	4.6	2.2	120

heat; for example, 30 minutes at 260 C decreases the strength of the steel in Table 3.1 to 75% of its 20 C value. The other wires are less affected, and tungsten is particularly stable in this respect. Steel wires cost about $1/kg, and other metal wires can be quite expensive, i.e. as much as $10/kg or more.

3.3. Inorganic Fibres and Whiskers

Inorganic materials in fibrous form have been widely used for many years. Asbestos retains its properties up to quite high temperatures, and has been widely used as an insulating material, and more recently, as a reinforcement for polymers. Glass fibres are also used for insulation, at moderate temperatures. They also provide very useful reinforcement for polymers, and this so-called "fibreglass' is nowadays a very commonly used material.

Other inorganic fibrous materials have been developed recently, most notably boron, and carbon. In addition, the potentialities of inorganic whiskers have also been appreciated in the last few years, and methods of making them in reasonable quantities for industrial use are being intensively developed. The various types of material will be discussed in turn.

3.3.1. Asbestos

There are two main types of asbestos—chrysotile, which consists of single crystal fibrils of 20 nm diameter, which are relatively easily separated from one another (Fig. 3.5), and amphiboles which are three-dimensional crystalline structures, but which have planes of easy cleavage, so that they can be broken down into fibres. Both forms are minerals, and are obtained by mining. Canada is one of the major sources.

Most of the asbestos that is used commercially is chrysotile, which consists of a three-layer sheet, each layer consisting of a plane of Mg^{++} with OH^- radicals, linked to a plane consisting of silicon and oxygen atoms. There is some mismatch between the planes, and this is thought to cause the rolling up of the plane to produce the material's "swiss roll" structure. The chemical formula for chrysotile is $3MgO.2SiO_2.2H_2O$. It is the lighter form of asbestos, is quite strong, and although difficult to separate into individual fibrils, is relatively easily separated into fibres of 10 microns in diameter. The fibres are normally damaged in the process of mining and preparing the material for use, but carefully produced material contains strong fibres with lengths of several centimetres (Table 3.2). It loses strength drastically 500 C.

Amphibole types of asbestos are less temperature resistant, losing much of their strength at 300 C. They are basically silicate structures containing oxides of other materials incorporated into the structure. For example, crocidolite has the chemical formula $(Na_2O.Fe_2O_3). 3FeO.8SiO_2.H_2O$. Other amphiboles contain different amounts of oxides of Fe, Mg, Ca, and Na. The moduli are a little higher than chrysotile, but they are usually not quite so strong (Table 3.2).

3.3.2. Glass

Glass is a word which covers a wide range of materials, usually containing more than

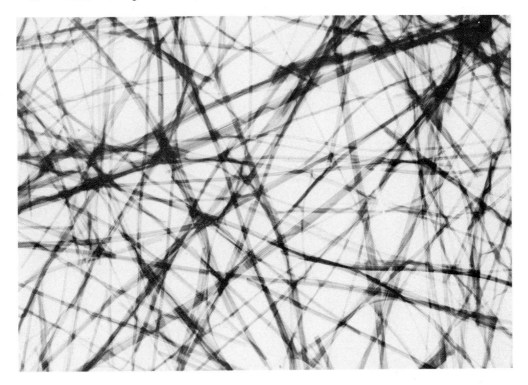

Fig. 3.5. Chrysotile asbestos. The fibres, of order 10-μm diameter, consist of many thousands of fibrils. (Courtesy of Woodhams, R. T., University of Toronto.)

50 % of silica (SiO_2) with random structures. They are often regarded as super-cooled liquids, and this state is referred to as the vitreous state. Ordinary glass, used for windows and bottles, typically contains 14% Na_2O, 10% CaO, 2.5% MgO, 0.6% Al_2O_3 and 0.7% SO_3, the remainder being SiO_2. It is relatively weak and has a high coefficient of thermal expansion compared with pure silica, coupled with very low thermal conductivity. It is very brittle and it is easily broken into small pieces by rapid cooling. It softens at about 700 C and so is relatively easy to form, though care has to be taken to relieve stresses developed during the forming and cooling processes. Fibres are made from this glass, but they do not have very good properties, being comparatively weak and relatively easily attacked by water.

Pyrex glass contains a relatively large amount of B_2O_3 (typically 12.9%) together with Na_2O (3.8%), Al_2O_3 (2.2%) and K_2O (0.5%), is much stronger than soda-lime glass, more resistant to water and chemicals, and has a much lower expansion coefficient. However, it is not easily drawn into fibres. The glass fibres used for textiles are of a different composition, and are the same as those used for reinforced polymers.

For reinforcement purposes, one of the most commonly used glasses is E-glass, developed originally for its good electrical properties. A typical composition is CaO 17.5%, Al_2O_3 14.4%, B_2O_3 8%, MgO 4.5%, K_2O 0.5%, Fe_2O_3 0.4%, F_2 0.3%, the remainder (54.4%) being SiO_2.

The fibres are usually made by melting and stirring the ingredients, then allowing the liquid to fall through holes 1–2 mm in diameter in a heated platinum plate. The glass is pulled away rapidly to draw the fibres down to about 10 microns diameter (Fig. 3.6). The platinum plate contains several hundred holes, and the fibres are all drawn together. To obtain strong fibres it is essential that the fibre surfaces do not touch anything, even another fibre. Consequently, they are coated, before being drawn together, with a "sizing". This is usually a starch-oil emulsion, or alternatively a special coating to ensure good adhesion between fibre and matrix when the fibres are later incorporated in a polymer.

A glass which was developed specifically for high strength and modulus is called S-glass. It retains its strength better at high temperatures than does E-glass, as well as having better properties at room temperature (Table 3.2). The major constituents of this glass are Al_2O_3 (25%) and MgO (10%) together with SiO_2. Both S-glass and E-glass are weakened by water, but S-glass is more resistant to acids, and less resistant to strong alkaline solutions.

The strongest known glass fibres are made from pure silica. At liquid nitrogen temperatures they can have a strength of 9.6 GPa and break at a strain of nearly 14%. This strength is not affected by water very much at all, and the fibres are very resistant to acids (apart from HF, and H_3PO_4 at high temperatures) but they are weakened by alkaline solutions.

All glass fibres are extremely sensitive to surface damage. Merely touching one fibre against another is sufficient to cause a crack which can reduce the strength to less than a half of the undamaged value. Handling the fibres is only possible if they have a protective layer on them, and even then considerable reduction in strength can occur unless great care is taken.

Glass fibres are relatively cheap, about $1/kg.

3.3.3. Boron

Boron fibres have been developed quite recently for aerospace applications. Boron has a very low density (2.3 Mgm^{-3}) and great potential strength, but is a very hard and brittle material, not suitable for drawing down into the form of fine wire. It is, therefore, made by deposition onto a very fine tungsten wire (12.5 microns diameter) using the chemical reaction $2BCl_3 + 3H_2 \rightarrow 2B + 6HCl$. Both reactants are gases, and other boron halides can be used instead of the chloride. A sketch of the apparatus used is shown in Fig. 3.7.

The tungsten wire is heated and strict temperature control is necessary to obtain consistent fibres having the optimum crystal size. Ideally, the crystals should be very small, about 2 to 3 nm in diameter, so that the material is practically amorphous. The best results are obtained at about 1100 C. The diameter of fibres normally produced is about 0.1 mm.

The fibres are extremely strong and stiff (Table 3.2), and retain their properties up to temperature of about 500 C. For use above this temperature they must be coated with SiC. This extends their usefulness up to 700 C, and also provides protection from reactive metallic matrices, such as Al and Ti. They are also very hard, so are not easily damaged by careless handling (unlike glass). However, they are extremely expensive to manufacture and cost several hundred dollars per kg. Attempts to reduce the cost by depositing the

The Fiber Glass Reinforcements Manufacturing Process

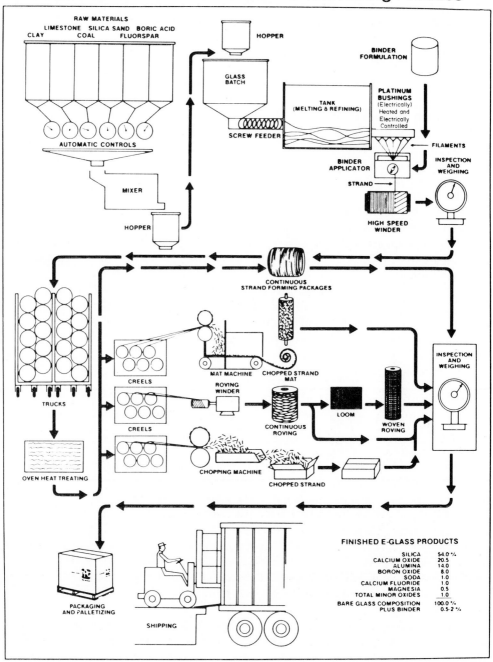

FIG. 3.6. Glass fibre production process. (Courtesy of PPG Industries Inc.)

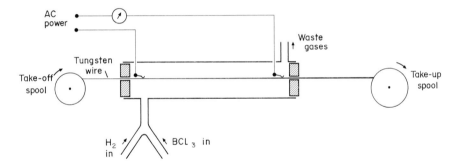

FIG. 3.7. Schematic drawing of chemical vapour deposition process used to make boron fibres.

boron onto a heated carbon fibre substrate have not been very successful; the fibres thus produced were weaker.

3.3.4. Silicon Carbide

These fibres are made using the same process as boron, namely, chemical vapour deposition, with a tungsten wire substrate. (A carbon fibre substrate has also been used, but the fibres produced are not so strong.) Chlorinated silanes are used, with hydrogen gas acting as a carrier. Decomposition occurs at high temperature to produce the silicon carbide. For example, with methyldichlorosilane, the reaction is $CH_3\,Si\,HCl_2$ $\rightarrow SiC + 2HCl + H_2$.

The fibres produced are light, stiff, and strong (Table 3.2) but are easily damaged by abrasion. They are usually made with diameters of 0.1 mm, but can be made much thicker. Although silicon carbide itself is quite strong up to 1400 C, it reacts with the tungsten substrate at much lower temperatures, and should not be used at temperatures above 900 C.

The cost of making the fibres is somewhat greater than boron, i.e. roughly $1000/kg.

Much finer fibres (about 10 μm dia.) can be made directly from carbon–silicon polymer fibres by pyrolysis and subsequent heat treatment. The process is similar to that used to produce carbon fibres, discussed in the next section.

3.3.5. Carbon

These are often called graphite fibres, and are usually made from polymer fibres by careful heat treatment. They can also be made from tar or bitumen.

Carbon can occur in three forms, diamond, graphite, and amorphous (or glassy). Only the crystalline forms have high modulus, and the only crystalline form of carbon that has been produced in fibre form resembles graphite. Amorphous carbon fibres can also be produced. They can be very strong (2.0 GPa) but their low modulus (70 GPa) renders them of little commercial interest so they will not be considered any further here.

The polymer fibres (for example rayon or polyacrylonitrile) are first heated at relatively low temperature (about 220 C) in air under tension to oxidize the fibres and stabilize them so that they can be heated at higher temperatures without disintegrating. They are then heated at temperatures up to 2500 C, usually under tension. After the initial conversion to carbon, some recrystallization occurs, so that the graphite planes are oriented along the fibre axis, as shown in Fig. 3.8.

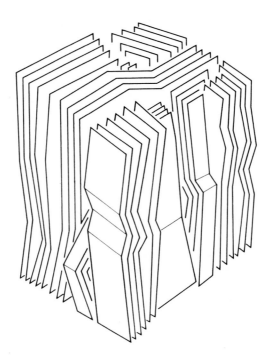

FIG. 3.8. Internal structure of a carbon fibre, showing the arrangement of planes of carbon atoms. (After Kelly, A. (1973) *Strong Solids* (2nd edn.), Oxford University Press, London.)

There is some advantage in having an oriented crystalline polymer as precursor, as this facilitates the production of high modulus fibre. Two types of fibre can be produced; a highly oriented high modulus type, and a less well oriented but slightly stronger variety (this requires less heating). Strengths and moduli are given in Table 3.2. The profile of the fibres is far from smooth, due in part to the profile of the polymer precursor (Fig. 3.9). Their diameter is about 8 μm. They are supplied as clusters of fibres, called tows.

Although the fibres retain their properties at temperatures over 2000 C in vacuum, they are very easily oxidized at temperatures above about 500 C.

The price of these fibres is being rapidly reduced, but it seems unlikely that they will ever cost less than $10/kg.

3.3.6. Sapphire Fibres

Continuous single crystal filaments can be grown from the melt under carefully controlled conditions. The sapphire is drawn upwards from a molten surface of Al_2O_3.

FIG. 3.9. An 8-micron diameter carbon fibre made from polyacrylonitrile. (Courtesy of Woodhams, R. T., University of Toronto.)

Although a wide range of cross-sections can be produced (depending on the geometry of the crucible lip) and many crystal orientations are possible, most filaments manufactured are circular, with a diameter of 0.25 mm, and are oriented so that the fibre axis is the hexagonal axis of symmetry of the crystal.

These fibres have excellent mechanical properties (Table 3.2) but soften at high temperatures due to dislocation mobility. Their strength decreases rapidly above 800 C. They are also very sensitive to abrasion.

They are extremely expensive; more than $1000/kg.

3.3.7. Inorganic Whiskers

These whiskers are the nearest approach we have to the perfect material. They are single crystals of about 1 micron in thickness, having lengths of up to 3 or 4 mm. They have hexagonal, square, or parallelogram sections, the faces being crystal faces determined by the internal arrangement of atoms in the structure. They can be grown with only one dislocation, a screw dislocation along the whisker axis, which cannot move under an axial tensile stress. The surfaces are close to being atomically smooth, and thus free from cracks, and under suitable conditions, using pure materials, the internal structure is free of inclusions and cracks, or other faults.

Whiskers can be made from a wide range of materials, and even alkali halides grown in this form have very high strengths. For example, whiskers of LiF have been produced which have a strength of 4.5 GPa.

The properties of the whiskers depend a great deal on growth conditions, surface perfection, and diameter. Thick whiskers have many dislocations in them and are relatively weak. Growth steps also occur on the surface of thick whiskers, and act as stress raisers, lowering the strength. One micron is the maximum diameter normally suitable.

To produce kilogram quantities of high quality whiskers is extremely difficult, and, so far, good whiskers have been very expensive (hundreds of dollars/kg). Their properties, however, far exceed those of the other forms of strong solids. Table 3.2 lists the properties of three types of whisker which resist high-temperature oxidation conditions. Whiskers produced in kilogram quantities, in commercial batch processes, do not come near to achieving the properties indicated in the table. Before being used in composites they have to be carefully sorted to remove large diameter fibres and very short ones. The final product then usually has to be aligned.

TABLE 3.2. *Representative Values for Strength, Modulus, and Maximum Use Temperature of Inorganic Fibres, Whiskers, and Planar Materials*
Representative values only are given. There is considerable variation in values reported for many of these materials. The temperatures are only intended as some indication of relative resistance to heating. In practice, maximum temperatures depend on stress and chemical environment.

Material	Density $(Mg\ m^{-3})$	Strength (GPa)	Young's modulus (GPa)	Temperature (°C)
Chrysotile asbestos	2.5	5.5	160	500
Amphibole asbestos	3.3	4.1	190	300
E-glass	2.54	3.4	72	550
S-glass	2.48	4.8	85	650
Fused silica	2.2	5.8	72	750
Boron	2.6	3.5	420	700 [‡]
Graphite (stiff)	1.9	2.3	377	2500 [‡]
Graphite (strong)	1.8	2.8	233	2000 [‡]
Al_2O_3 fibres	4.0	2.0	470	800
SiC fibres	3.4	2.3	480	900
Al_2O_3 whiskers	4.0	15	2250 [†]	1200
SiC whiskers	3.2	21	840 [†]	1600
BeO whiskers	3.0	6.9	720 [†]	1500
SiC platelets	3.2	10	480	1600
AlB_2 platelets	~2.7	~6	~500	>1000
Mica platelets	2.8	3.1 [§]	226	400

[†] Maximum value, with most favourable crystallographic orientation.
[‡] 500 C in oxidizing atmosphere.
[§] Maximum value, with perfect edges. In practice, strength of small mica flakes ~0.85 GPa.

3.3.8. Thermal Expansion

The thermal expansion coefficients of fibres are very important for reinforcement. Very often stress transfer between fibres and matrix requires that the matrix shrinks onto the fibres during manufacture. Since composites are usually made at elevated temperatures this requires that the matrix thermal expansion coefficient is larger than that of the fibres. Unfortunately, however, differential thermal expansions can cause problems when composites are thermally cycled. These effects are discussed later.

Table 3.3 gives some thermal expansion coefficients for fibres at about 100 C. The coefficients for matrices are presented in the appropriate places in Chapters 9, 10, 11.

TABLE 3.3. *Coefficients of Thermal Expansion of Fibres*

Fibre	$\alpha_f (\mu K^{-1})$
Alumina	6.2 – 6.8
Asbestos	9.2
Beryllium	12
Boron	8.3
Carbon	8.0[†]
E-glass	15.5
Kevlar	59[†]
Molybdenum	5.0
S-glass	8.9
Silicon carbide	4.8
Steel	12
Tungsten	4.3

† Radical expansion coefficient. Axial coefficients are $0.5 \ \mu K^{-1}$ for carbon and $-2 \mu K^{-1}$ for Kevlar.

3.4. Single Crystal Platelets

Many materials grow naturally in platelet form, and though platelets are not normally as strong as whiskers, they can still have strengths of up to 10 GPa. Figure 3.3 indicates the relative strengths of platelets and whiskers of SiC, compared with more equiaxed crystals. There is, however, tremendous scatter in the values, especially for platelets.

Silicon carbide platelets have edges which are parallel to important directions of the crystal. Thus they have hexagonal or equilateral triangular shapes. Mica, on the other hand, is produced in irregular shapes (Fig. 3.10), due to the grinding it has been subjected to after mining. The imperfections of the edges cause the strength of platelets to be much less than the potential strength of the material. In composites, for example, mica appears to break at 0.85 GPa, compared with a potential strength of 2.7 GPa for mica not weakened by edge effects. (The high value was obtained by stressing only the centre of a large mica sheet.)

The properties of a number of materials that can be obtained in planar form are given in Table 3.2. Planar materials have the advantage that, when suitably oriented, they can stiffen a material in two dimensions rather than only one, as in the case of aligned fibres. However, they cannot impart as much strength to the composite as one might expect, owing to interactions between nearest neighbour platelets.

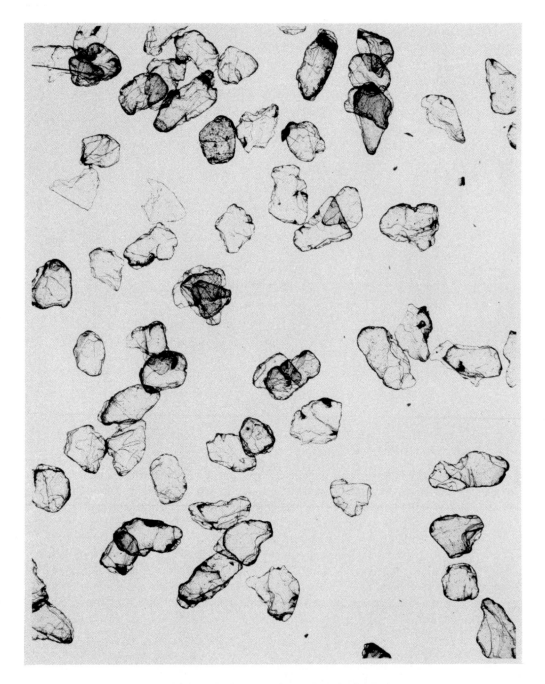

FIG. 3.10. Mica platelets, about 100 μr   in diameter.
(Courtesy of Woodhams, R. T., University of Toronto.)

3.5. Forms of Reinforcement

In order to use the high strength of fibrous materials on a commercial scale, it is necessary to handle very large numbers of them at a time. In the case of glass fibres, a one millimetre diameter rod composed only of fibres tightly packed, would contain 10^4 fibres. Such rods, if made of chrysotile asbestos, would contain 10^9 fibrils. Methods have been developed over a long period for the routine handling of large numbers of individual fibres.

The art of doing this was originally developed for the handling of textile fibres for the manufacture of clothing and of cellulose fibres for making paper and cardboard. These methods have recently been adapted for the handling of asbestos and glass, and very recently for the handling of single crystal whiskers.

3.5.1. Random Mat

This is the sort of arrangement normally present in paper and cardboard. Single crystal whiskers, which grow as felt-like small cushions, of very low density, may be beaten down

Fig. 3.11. Fibreglass roving. (Courtesy of Fiberglas Canada Ltd.)

FIG. 3.12. Fibreglass yarn. (Courtesy of Fiberglas Canada Ltd.)

into a mat mechanically, with some loss in aspect ratio (length/diameter) of the individual crystals. Asbestos fibres, after being broken up, can be settled into random mats. Glass fibres for use in reinforcement is manufactured in multiple, continuous fibre strands called roving—see Fig. 3.11., or twisted yarn, Fig. 3.12. To produce random arrangements the rovings are chopped into short lengths (3–50 mm) and allowed to settle on a sheet. Fig. 3.13 shows an example of chopped strand mat of fibreglass.

Mats to be used for reinforcement are often made with some polymer present to hold the fibres in place. They are an inefficient form of reinforcement because it is impossible to pack large numbers of straight fibres tightly unless they are all parallel. The maximum volume fraction that can be obtained when they are random in a plane is about 0.4. For three-dimensional random arrangements the packing efficiency is much worse, especially for fibres and whiskers having large aspect ratios.

Continuous fibres can also be laid into mats. This arrangement has the same reinforcing potential as chopped strands, but has different handling and moulding characteristics. It can be moulded into more complicated shapes, and does not tear so easily.

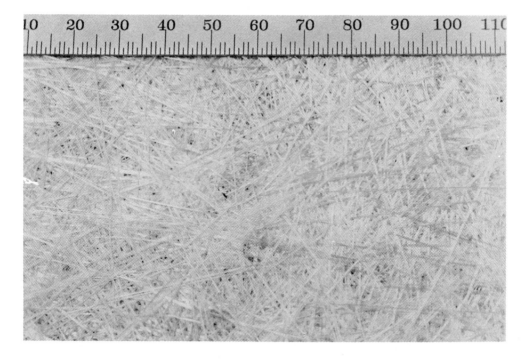

FIG. 3.13. Chopped strand mat; scale in mm. (Courtesy of Fiberglas Canada Ltd.)

3.5.2. Fabrics

Yarns can be produced from a wide range of fibres and whiskers. In the case of short fibres and whiskers, the yarn must be spun (or twisted) to hold the fibres together. Continuous fibres require no spinning, but it is often advantageous to do so. Fabrics are produced from these yarns by normal weaving processes. Figure 3.14 shows a typical glass fabric. If the fibres are not spun the fabrics are usually denser, Fig. 3.15, and involve much less fibre flexure.

Since the fibres are highly flexed in fabrics made by conventional weaving processes, some fibre straightening can occur when the fabric is stretched. This results in a relatively low modulus composite, with similar properties in all directions. The modulus may be improved in one direction, with consequent loss of modulus and strength in the direction at right angles to it, if the fabric is made with the fibres completely straight in one direction, and with only a few fine threads in the other direction. This is a particularly useful method of weaving for very stiff fibres which can withstand relatively little flexure, for example, graphite and boron. Owing to their large diameters, boron fibres are extremely difficult to weave.

Woven fabrics should be used when high shear strengths are required in the plane of the reinforcing sheet (for example, when the sheet is subject to flexure). The more unidirectional weaves generally have lower shear strengths than conventional weaves.

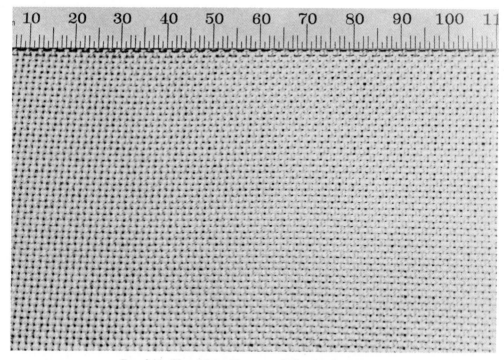

FIG. 3.14. Glass fabric. (Courtesy of Fiberglas Canada Ltd.)

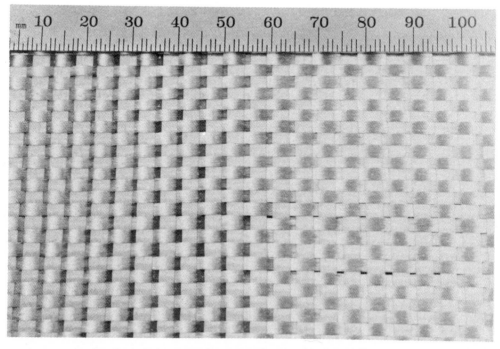

FIG. 3.15. Non-twisted glass fabric. (Courtesy of Fiberglas Canada Ltd.)

3.5.3. Tapes

Glass, boron, steel, Kevlar, and carbon fibres are produced in tape form. Some type of binder is required to retain the shape. For polymer reinforced materials this is a resin which is partly cured. For reinforced metals, metal powders or foils are included in the tape, and the binder is a material which produces only volatile products on heating, so that only metal and fibres are present in the final hot-pressed composite. Methods have also recently been developed for the production of sheets of aligned short fibres.

Paper-like tapes or sheets can also be produced with materials like mica. Again, some binder has to be present to hold the tape in shape.

Further Reading

BROUTMAN, L. J. and KROCK, R. H. *Modern Composite Materials*, (Addison-Wesley, Reading, Mass, 1967).
LUBIN, G. (ed.) *Handbook of Fiberglass and Advanced Plastics Composites* (Van Nostrand Reinhold, New York, 1969), Chapters 7–12.
PARRATT, N. J. *Composite Materials Technology* (Van Nostrand Reinhold, London, 1972).

4

Composite Mechanics

IN THIS chapter we will describe a simple slip theory of reinforcement which gives the modulus and strength, in the fibre direction, for aligned short-fibre composites. The advantages of continuous fibres are thereby demonstrated. The rest of the chapter deals with the properties of continuous fibre composites.

One of the most useful forms of composite for the construction of high-performance structural elements for aerospace is the lamina, made from aligned fibre tapes, containing also partly polymerized matrix (or metal powder). The tapes are also made with woven fibres; these have inferior stiffness, and cannot have such high fibre volume fractions, but are generally much tougher than aligned fibre tapes. The structure is made by gluing the laminae together with the fibres aligned in directions best suited to the stresses to be encountered. This is a laminate. This chapter will briefly describe the properties of laminae and laminates. (The reader interested in the design of laminates should refer to more specialist texts. See the reading list at the end of the chapter.)

A complete list of the symbols used in this and other chapters is given in the Appendix.

4.1. Reinforcement by Slip

We will consider straight fibres that are perfectly elastic up to their breaking-points, σ_{fu}. (This is true of non-metal, non-polymer fibres such as boron, glass, and carbon etc.). Their tensile stresses are governed by an interfacial shear stress parallel to the fibre surface, which we will assume has a constant value, τ_i, near the ends, and is zero near the centre, Fig. 4.1.

The matrix is assumed to be either elastic-perfectly plastic (for metals) or entirely elastic (for polymers).

Consider the fibre end region, CD. The fibre stress will change from σ_f to $\sigma_f + d\sigma_f$ along an element of length dx (Fig. 4.2). For the surface shear forces to be in equilibrium with the tensile forces in the fibre

$$\pi r^2 d\sigma_f = -2\pi r \, dx \, \tau_i$$

where $2r$ is the fibre diameter. Rearranging and simplifying

$$\frac{d\sigma_f}{dx} = -\frac{2\tau_i}{r}. \tag{4.1}$$

If $\sigma_f = 0$ at $x = L$ this integrates to give

$$\sigma_f = \frac{2\tau_i}{r}(L - x). \tag{4.2}$$

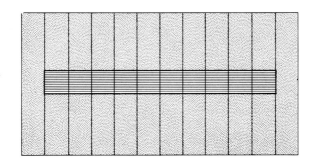

(a)

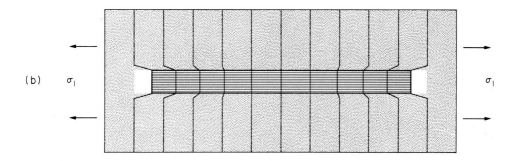

(b)

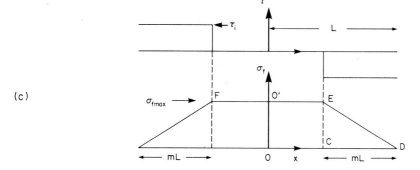

(c)

FIG. 4.1. Single fibre composite element: (a) unstressed and (b) stressed. (c) shows the fibre–matrix interfacial stress and the fibre internal stress for stress transfer by slip.

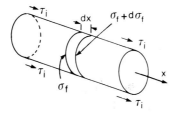

FIG. 4.2. Short length of fibre.

In the centre section, OC, where $x \leqslant L(1-m)$ the stress is constant, i.e. $\sigma_f = \sigma_{f\max}$ where

$$\sigma_{f\max} = 2ms\tau_i \tag{4.3}$$

and s is the aspect ratio, i.e.

$$s = L/r. \tag{4.4}$$

In the composite there are many fibres. Consider a composite containing fibres which all have the same length and diameter, and are all parallel, but are otherwise randomly positioned, as shown in Fig. 4.3. Then any section normal to the bar axis, MN for example, will intersect fibres at all possible positions along their lengths, so long as there are a very large number of them in the cross-section.

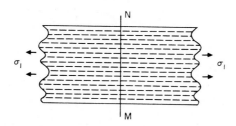

FIG. 4.3. AA′ represents a random cross-section in an aligned fibre composite.

The load carried by the fibres will be the total fibre area across the bar section, multiplied by the average fibre stress. The area fraction is equal to the volume fraction in any random cross section; consequently the load carried by the fibres is $AV_f\bar{\sigma}_f$ where A is the area of the cross section, V_f is the volume fraction and $\bar{\sigma}_f$ the average stress of the fibres. Similarly, the load carried by the matrix is $AV_m\bar{\sigma}_m$, the subscript m referring to the matrix.

The total load carried by the composite is $A\sigma_1$, the fibres being aligned in the 1 direction. Consequently the composite obeys the RULE OF AVERAGES.

$$\sigma_1 = V_f\bar{\sigma}_f + V_m\bar{\sigma}_m. \tag{4.5}$$

The average fibre stress is the area under the line O′ED (Fig. 4.1) divided by the half length of the fibre, i.e.

$$\bar{\sigma}_f = \sigma_{f\max}(1-m/2)$$

so that the composite stress comes to

$$\sigma_1 = V_f\sigma_{f\max}(1-m/2) + V_m\bar{\sigma}_m. \tag{4.6}$$

Since no slip takes place there, the fibres and matrix strains in the centre region (OC, Fig. 4.1) are the same. We neglect the effect of the disturbances at the fibre end on the tensile strain of the matrix. We can therefore equate matrix and composite strains, and write

$$\varepsilon_1 = \sigma_{f\max}/E_f \tag{4.7}$$

and, for a matrix below the yield point (i.e. $\varepsilon_1 < \varepsilon_{my}$),

$$\bar{\sigma}_m = E_m\varepsilon_1. \tag{4.8}$$

We can now use equation (4.3) for m, and substitute equations (4.7) and (4.8) into equation (4.6) to get the stress-strain relation

$$\sigma_1 = (V_f E_f + V_m E_m)\varepsilon_1 - \frac{V_f E_f^2 \varepsilon_1^2}{4\tau_i s} \qquad (4.9)$$

so long as $m \leqslant 1$; i.e. using equations (4.3) and (4.7), $\varepsilon_1 < 2s\tau_i/E_f$. Let ε_{1p} be the limiting value of ε_1. Thus

$$\varepsilon_{1p} = 2s\tau_i/E_f. \qquad (4.10)$$

In addition σ_{fmax} must not exceed the fibre breaking-stress, σ_{fu}.

When $m = 1$ the fibre stress is no longer a function of ε_1. Instead, $\bar{\sigma}_f = \sigma_{fmax}/2$ where σ_{fmax} is given by equation (4.3) with $m = 1$. Thus our stress–strain relation becomes

$$\sigma_1 = V_f s\tau_i + V_m E_m \varepsilon_1. \qquad (4.11)$$

We can now, in principle, draw stress–strain curves. If we choose the combination of stiff carbon and polycarbonate we have a matrix which yields above the fibre breaking-strain. This polymer deforms a great deal at approximately constant stress (Fig. 1.2), and so a well-adhering fibre will develop a shear stress at the matrix interface which for present purposes is sufficiently accurately represented by $\tau_i = \sigma_{my}/2$ where σ_{my} is the matrix yield stress. Figure 4.4 shows theoretical curves for this material. They have the following features.

1. When the aspect ratio is very large the composite obeys the RULE OF MIXTURES for modulus. We write $E_1 = \sigma_1/\varepsilon_1$ and let $s \to \infty$ in equation (4.9). Thus

$$E_1 = V_f E_f + V_m E_m. \qquad (4.12)$$

The composite behaviour therefore appears to be entirely elastic, despite the fact that shear yielding in the matrix is taking place. The reason is, of course, that the volume of sheared material is negligibly small.

2. The material does not obey the RULE OF MIXTURES for strength. When $\varepsilon_1 = \varepsilon_{fu}$ $(= \sigma_{fu}/E_f)$ the fibres break (i.e. $\sigma_{fmax} = \sigma_{fu}$). The composite strength is then

$$\sigma_{1u} = V_f \sigma_{fu} + V_m E_m \sigma_{fu}/E_f \qquad (4.13)$$

for $s \to \infty$, and less than this for finite s.

3. For finite s the stress–strain trajectory is curved, i.e. there is loss of effective modulus as well as strength.

4. For aspect ratios below a critical value the composite does not necessarily fail at the fibre breaking-strain. The fibres are now too short to be stressed to the breaking-point. The critical aspect ratio is given by equation (4.3) with $m = 1$ and $\sigma_{fmax} = \sigma_{fu}$ i.e.

$$s_c = \sigma_{fu}/2\tau_i. \qquad (4.14)$$

Figure 4.5 shows fibres longer than, equal in length, and shorter than the critical length. At the critical length $s = s_c$. The fibres are stressed to the breaking-stress, where possible.

Employing the same steps used to develop equation 4.9 but with $\sigma_{fmax} = \sigma_{fu}$ we obtain for the composite strength, σ_{1u}, for $s \geqslant s_c$,

$$\sigma_{1u} = \sigma_{fu}\{V_f(1 - s_c/2s) + V_m E_m/E_f\} \qquad (4.15)$$

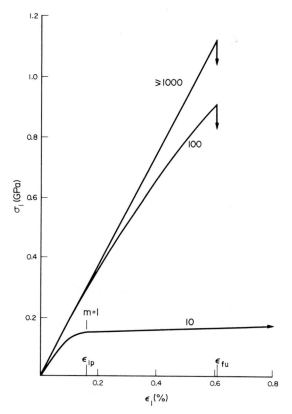

FIG. 4.4. Theoretical stress–strain curves for stiff carbon-polycarbonate with $\tau_i = \sigma_{my}/2$, for the various aspect ratios marked on the curves. Stress transfer by slip. $V_f = 0.55$.

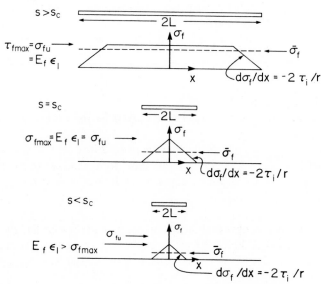

FIG. 4.5. Three fibres of different lengths, stressed to the maximum possible by surface shear stress τ_i.

while for $s < s_c$ it is given by

$$\sigma_{1u} = V_f s \tau_i + V_m \sigma_{my} \tag{4.16}$$

where we have assumed failure when $E_m \varepsilon_1 = \sigma_{my}$ (equation (4.11)).

In most reinforced polymers it is believed that frictional slip generates the interfacial shear stresses i.e.

$$\tau_i = \mu \sigma_r \tag{4.17}$$

where μ is the coefficient of friction and σ_r is the normal stress at the fibre–matrix interface. Since the coefficient of friction can be in the region 0.2 to 0.3 and the normal stress, which results mainly from cure shrinkage of the matrix, can be only 20–30 MPa, τ_i is often only a few MPa. Consequently, critical aspect ratios are correspondingly large (often an order of magnitude greater than those for stress transfer by matrix flow).

Figure 4.6 shows stress–strain curves, for the same aspect ratios, for E-glass-epoxy with $\tau_i = 6$ MPa. It can be seen that, for a given aspect ratio, deviations from the Rule of Mixtures are much greater than in the previous case.

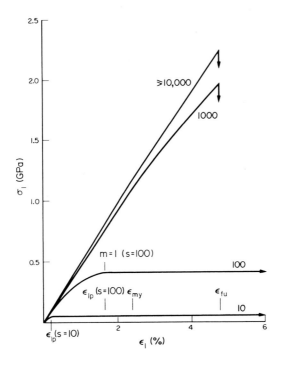

FIG. 4.6. Theoretical stress–strain curves for E-glass-epoxy with $\tau_i = 6$ MPa, for the various aspect ratios marked on the curves. Stress transfer by slip. $V_f = 0.67$.

The low modulus and high strength of the glass means that the yield strength of the matrix is exceeded before the fibres break. At these high volume fractions ($V_f = 0.67$) the direct effect of the relatively weak and low modulus matrix is so small that effects due to this are insignificant. The matrix merely serves as a stress transfer medium.

It is a different matter when we come to reinforced metals, though. When reinforced metals are made with great care, slip usually takes place at a stress close to the shear yield stress of the metal. Thus,

$$\tau_i \simeq \sigma_{my}/2$$

so that

$$S_c = \sigma_{fu}/\sigma_{my}. \tag{4.18}$$

Usually the metal is relatively pure and well annealed, and hence quite weak (e.g. for silica–aluminium $\sigma_{my} \simeq 60$ MPa). Consequently critical aspect ratios usually lie in the range 40–80.

The softness of the metal matrix introduces a new feature into the stress–strain curve. ε_{my} is usually less than ε_{fu}, i.e. the metal yields before the fibres break. Because of this we get a knee in the stress–strain curve at $\varepsilon_1 = \varepsilon_{my}$, as can be seen in Fig. 4.7.

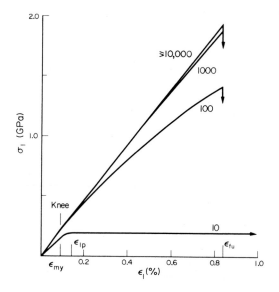

FIG. 4.7. Theoretical stress–strain curves for B–Al with $\tau_i = \sigma_{mu}/2$, for the various aspect ratios marked on the curves. Stress transfer by slip. $V_f = 0.48$.

Below the knee ($\varepsilon_1 < \varepsilon_{my}$) σ_1 is given by equation 4.9. Above the knee the matrix has yielded, so $\bar{\sigma}_m$ is given by σ_{my} rather than $\varepsilon_1 E_m$, and instead of equation (4.9) we have

$$\sigma_1 = V_f E_f \varepsilon_1 \left(1 - \frac{E_f \varepsilon_1}{2\sigma_{my} S}\right) + V_m \sigma_{my}. \tag{4.19}$$

Note that we have replaced $4\tau_i$ by $2\sigma_{my}$ in this equation.

Although continuous fibres give composites which obey the Rule of Mixtures for modulus below the knee, above the knee the slope of the stress–strain curve is reduced to $V_f E_f$.

However, the strength does obey the Rule of Mixtures for continuous fibres. Remember that we have assumed elastic-perfectly plastic behaviour. Thus $\sigma_{my} = \sigma_{mu}$. Putting this, and $\varepsilon_1 = \varepsilon_{fu}$ into equation 4.19 for $s \to \infty$ gives

$$\sigma_{1u} = V_f \sigma_{fu} + V_m \sigma_{mu}. \tag{4.20}$$

For smaller aspect ratios, so long as $s > s_c$

$$\sigma_{1u} = V_f \sigma_{fu}(1 - s_c/2s) + V_m \sigma_{mu} \tag{4.21}$$

Finally, for $s < s_c$ we use equation (4.16) with $2\tau_i$ and σ_{my} replaced by σ_{mu}

$$\sigma_{1u} = \sigma_{mu}(V_f s/2 + V_m). \tag{4.22}$$

Stress–strain curves are shown in Fig. 4.7.

Figures 4.6 and 4.7 clearly demonstrate the advantages of long fibres. Notice, however, that once $s > 50s_c$ the fibre length has negligible effect (i.e. $< 1\%$). For reinforced polymers, this requires that carbon or glass fibres have a length of at least 7 cm. For reinforced metals they must be more than 2 cm.

More detailed analysis of E_1 for continuous aligned fibres show that the fibres introduce additional constraints by virtue of their Poisson's shrinkage. This increases the modulus, so that it should be a little greater than the Rule of Mixtures. Such effects have been observed with very carefully made reinforced polymers.

4.2. Transverse Elastic Properties

The calculation of elastic moduli, apart from the Young's modulus in the fibre direction, is presently surrounded by controversy, because of disagreements over the approximations needed. This matter is discussed at some length by Jones, and the interested student should refer to the reading list at the end of this chapter.

In this description we will present the most simple theory. This requires an assumption that either the stresses or the strains in both components are the same. This is only true of models in which the fibres are lumped together in a long rectangular prism, with matrix material attached to two opposite sides of it. (It is this rather drastic assumption about the contiguity that is the main bone of contention among theorists.) The fibres are also assumed to have $s \gg s_c$.

4.2.1. Transverse Young's Moduli

Figure 4.8a represents a view of an aligned fibre composite, stressed transversely to the fibres. For the analysis we lump all the fibres together, in a band normal to the stress, Fig. 4.8b. Consequently, the stress in fibres and matrix is the same, so that the strains are $\varepsilon_f = \sigma_2/E_f$ and $\varepsilon_m = \sigma_2/E_m$. The total displacement, $\varepsilon_2 t$, is the sum of fibre and matrix displacements $V_f \varepsilon_f t$ and $V_m \varepsilon_m t$, so that

$$\varepsilon_2 = V_f \varepsilon_f + V_m \varepsilon_m$$

and since $E_2 = \sigma_2/\varepsilon_2$ we can substitute for ε_2, ε_f and ε_m in the above equation and obtain

$$\frac{1}{E_2} = \frac{V_f}{E_f} + \frac{V_m}{E_m} \tag{4.23}$$

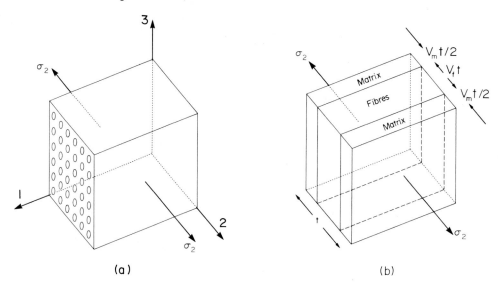

FIG. 4.8. (a) Axes used for composites. The fibres are in the 1 direction, and the applied stress is in the 2 direction. (b) Lumped fibres and matrix.

The same equation is obtained for E_3.

We can easily extend this treatment to the case of planar random fibres. Such composites will have moduli normal to the plane which are also given by equation (4.23).

In the case of most reinforced polymers $E_f \gg E_m$ and so $E_2 = E_3 \simeq E_m / V_m$. Since the fibre modulus has so little effect, the nature of the contiguity assumption is not too critical.

4.2.2. Shear Moduli

As in the previous case we assume that the fibres and matrix experience the same stress, this time the shear stress τ_{12} (Fig. 4.9). The total matrix displacement, u_m, is equal to $V_m t \gamma_m$ and the total fibre displacement, u_f, is $V_f t \gamma_f$ where γ's are the appropriate shear strains. Thus

$$\gamma_{12} = \frac{u_m + u_f}{t} = V_f \gamma_f + V_m \gamma_m.$$

But $\gamma_f = \tau_{12}/G_f$, $\gamma_m = \tau_{12}/G_m$ and $\gamma_{12} = \tau_{12}/G_{12}$, so that

$$\frac{1}{G_{12}} = \frac{V_f}{G_f} + \frac{V_m}{G_m}. \tag{4.24}$$

The same equation is obtained for G_{13}, and for in-plane shears with planar-random fibre reinforced materials. Once again, the nature of the contiguity assumption is not important for reinforced polymers with $G_f \gg G_m$.

This type of analysis cannot be used to estimate G_{23}.

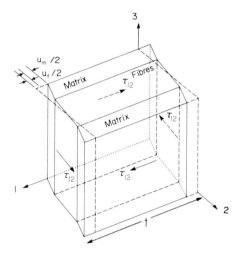

FIG. 4.9. Composite under shear stress τ_{12}.

4.2.3. Poisson's Ratios

When the composite is subject only to a stress σ_1 (Fig. 4.10), the matrix will contract, normal to the stress, a total amount $u_m = v_f \varepsilon_1 V_m t$ and the fibres by a total amount $u_f = v_f \varepsilon_1 V_f t$. The total contraction, u_2, is the sum of u_f and u_m, and the corresponding composite strain, ε_2 is $= u_2/t$. Thus

$$\varepsilon_2 = -(V_f v_f + V_m v_m)\varepsilon_1$$

and since $v_{12} = -\varepsilon_2/\varepsilon_1$

$$v_{12} = V_f v_f + V_m v_m. \tag{4.25}$$

By symmetry $v_{12} = v_{13}$, and v_{21} and v_{31} can be obtained using equation (1.50).

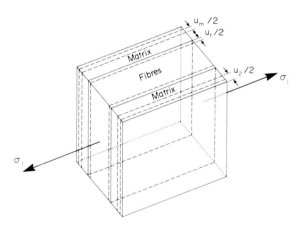

FIG. 4.10. Composite showing Poisson's shrinkage in the 2 direction due to stress in the 1 direction.

This simple analysis cannot be used to obtain v_{23}, nor can it be used for planar random fibres.

4.3. Laminae

Figure 4.11 shows an aligned fibre lamina. Such laminae are the building blocks used to make high-performance structural elements. An understanding of their properties is essential if we are to analyse structures made from them.

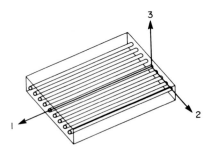

FIG. 4.11. Axes used for aligned fibre lamina.

4.3.1. Stress–Strain Relations

For laminae we can assume that plane stress conditions apply. Thus half the terms in our equations relating stresses to strains (equations (1.43) to (1.51)) are zero. Further simplification results because the laminae are orthotropic. In Fig. 4.11 the fibres are in the 1 direction. The plane stress state is defined by

$$\sigma_3 = \tau_{23} = \tau_{31} = 0.$$

Strains are still present normal to the plane of the lamina:

$$\varepsilon_3 = S_{13}\sigma_1 + S_{23}\sigma_2.$$

The in-plane strain–stress relations become

$$\begin{vmatrix} \varepsilon_1 \\ \varepsilon_2 \\ \gamma_{12} \end{vmatrix} = \begin{vmatrix} S_{11} & S_{12} & 0 \\ S_{12} & S_{22} & 0 \\ 0 & 0 & S_{66} \end{vmatrix} \cdot \begin{vmatrix} \sigma_1 \\ \sigma_2 \\ \tau_{12} \end{vmatrix}. \tag{4.26}$$

Engineering constants can be used in place of the compliances S_{11}, S_{12}, S_{22}, and S_{66}. They are given in equation (1.51). The strain–stress relations can be inverted using reduced stiffness, Q_{ij}.

$$\begin{vmatrix} \sigma_1 \\ \sigma_2 \\ \tau_{12} \end{vmatrix} = \begin{vmatrix} Q_{11} & Q_{12} & 0 \\ Q_{12} & Q_{22} & 0 \\ 0 & 0 & Q_{66} \end{vmatrix} \cdot \begin{vmatrix} \varepsilon_1 \\ \varepsilon_2 \\ \gamma_{12} \end{vmatrix} \tag{4.27}$$

where

$$Q_{11} = \frac{S_{22}}{S_{11}S_{22} - S_{12}^2} = \frac{E_1}{1 - v_{12}v_{21}} \qquad (4.28)$$

$$Q_{12} = \frac{S_{12}}{S_{11}S_{12} - S_{12}^2} = \frac{v_{12}E_2}{1 - v_{12}v_{21}} \qquad (4.29)$$

$$Q_{22} = \frac{S_{11}}{S_{11}S_{22} - S_{12}^2} = \frac{E_2}{1 - v_{12}v_{21}} \qquad (4.30)$$

$$Q_{66} = 1/S_{66} = G_{12}. \qquad (4.31)$$

In practice $v_{12}v_{21}$ is usually very small and can be neglected $(v_{12}v_{21} = v_{12}^2 E_2/E_1)$.

4.3.2. Oblique Stress–Strain Relations

To determine the stress–strain relations in directions other than the 1 and 2 directions (and hence the moduli in other directions) we use equations (1.8) and (1.10), together with the analogue of equation (1.8) for the stress parallel to the plane shown in Fig. 1.12.

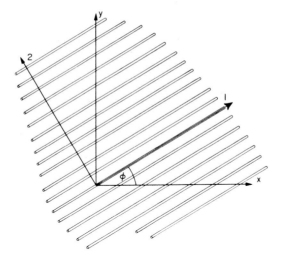

Fig. 4.12. Laminate axes (x, y) and lamina axes $(1, 2)$.

We will use our original x and y axes in Fig. 1.13 for the 1 and 2 axes, and use the direction of σ as our new x axis, as shown in Fig. 4.12. Thus instead of equation (1.8) we have

$$\sigma_x = \sigma_1 \cos^2 \phi + \sigma_2 \sin^2 \phi - 2\tau_{12} \sin \phi \cos \phi$$

and instead of equation (1.10) we have

$$\tau_{xy} = (\sigma_1 - \sigma_2) \sin \phi \cos \phi + \tau_{xy}(\cos^2 \phi - \sin^2 \phi).$$

We obtain σ_y by considering a plane at right angles to the diagonal plane in Fig. 1.12. Thus

$$\sigma_y = \sigma_1 \sin^2 \phi + \sigma_2 \cos^2 \phi + 2\tau_{12} \sin \phi \cos \phi.$$

These three equations can be written in matrix form:

$$\begin{vmatrix} \sigma_x \\ \sigma_y \\ \tau_{xy} \end{vmatrix} = \begin{vmatrix} \cos^2 \phi & \sin^2 \phi & -2 \sin \phi \cos \phi \\ \sin^2 \phi & \cos^2 \phi & 2 \sin \phi \cos \phi \\ \sin \phi \cos \phi & -\sin \phi \cos \phi & \cos^2 \phi - \sin^2 \phi \end{vmatrix} \cdot \begin{vmatrix} \sigma_1 \\ \sigma_2 \\ \tau_{12} \end{vmatrix}. \tag{4.32}$$

If we write our transformation matrix as $[T]$, we find that the same transformation applies to the strains ε_x, ε_y, and $\frac{1}{2}\gamma_{xy}$, with respect to the strains ε_1, ε_2, and $\frac{1}{2}\gamma_{12}$.

$$\begin{vmatrix} \varepsilon_x \\ \varepsilon_y \\ \frac{1}{2}\gamma_{xy} \end{vmatrix} = [T] \cdot \begin{vmatrix} \varepsilon_1 \\ \varepsilon_2 \\ \frac{1}{2}\gamma_{12} \end{vmatrix} \tag{4.33}$$

The stress–strain relationships for these axes are

$$\begin{vmatrix} \varepsilon_x \\ \varepsilon_y \\ \gamma_{xy} \end{vmatrix} = \begin{vmatrix} \bar{S}_{11} & \bar{S}_{12} & \bar{S}_{16} \\ \bar{S}_{12} & \bar{S}_{22} & \bar{S}_{26} \\ \bar{S}_{16} & \bar{S}_{26} & \bar{S}_{66} \end{vmatrix} \cdot \begin{vmatrix} \sigma_x \\ \sigma_y \\ \tau_{xy} \end{vmatrix} \tag{4.34}$$

where $[\bar{S}]$ is the transformed compliance matrix, which can be evaluated by suitable matrix manipulation to give

$$\bar{S}_{11} = S_{11} \cos^4 \phi + (2S_{12} + S_{66}) \sin^2 \phi \cos^2 \phi + S_{22} \sin^4 \phi \tag{4.35}$$

$$\bar{S}_{12} = S_{12} (\sin^4 \phi + \cos^4 \phi) + (S_{11} + S_{22} - S_{66}) \sin^2 \phi \cos^2 \phi \tag{4.36}$$

$$\bar{S}_{22} = S_{11} \sin^4 \phi + (2S_{12} + S_{66}) \sin^2 \phi \cos^2 \phi + S_{22} \cos^4 \phi \tag{4.37}$$

$$\bar{S}_{16} = (2S_{11} - 2S_{12} - S_{66}) \sin \phi \cos^3 \phi - (2S_{22} - 2S_{12} - S_{66}) \sin^3 \phi \cos \phi \tag{4.38}$$

$$\bar{S}_{26} = (2S_{11} - 2S_{12} - S_{66}) \sin^3 \phi \cos \phi - (2S_{22} - 2S_{12} - S_{66}) \sin \phi \cos^3 \phi \tag{4.39}$$

$$\bar{S}_{66} = 2(2S_{11} + 2S_{22} - 4S_{12} - S_{66}) \sin^2 \phi \cos^2 \phi + S_{66} (\sin^4 \phi + \cos^4 \phi). \tag{4.40}$$

Notice that we now have shear stress–tensile stress interactions, although these were not present when the axes were normal to the planes of symmetry in the structure.

We can write these expressions in terms of the engineering constants, using equation (1.51). For example

$$\frac{1}{E_x} = \frac{\cos^4 \phi}{E_1} + \left(\frac{1}{G_{12}} - \frac{2\nu_{12}}{E_1} \right) \sin^2 \phi \cos^2 \phi + \frac{\sin^4 \phi}{E_2}. \tag{4.41}$$

We have two extra engineering constants, $\eta_{xyx} = \bar{S}_{16} E_x$ which can be evaluated using equations (4.38) and (1.51), and η_{xyy}. Notice that if we express them as functions of ϕ, $E_x(\phi) = E_y(\pi/2 - \phi)$ and $\eta_{xyx}(\phi) = \eta_{xyy}(\pi/2 - \phi)$.

The engineering strain–stress relations are

$$\begin{vmatrix} \varepsilon_x \\ \varepsilon_y \\ \gamma_{xy} \end{vmatrix} = \begin{vmatrix} 1/E_x & -\nu_{xy}/E_x & \eta_{xyx}/E_x \\ -\nu_{xy}/E_x & 1/E_y & \eta_{xyy}/E_y \\ \eta_{xyx}/E_x & \eta_{xyy}/E_y & 1/G_{xy} \end{vmatrix} \cdot \begin{vmatrix} \sigma_x \\ \sigma_y \\ \tau_{xy} \end{vmatrix} \tag{4.42}$$

Also present are out-of-plane strains: ε_3 (this is present when $\phi = 0$) and γ_{13} and γ_{23} (only present when $\phi \neq 0$).

Figure 4.13 shows the elastic constants E_x and G_{xy} as a function of ϕ for E-glass-epoxy with $V_f = 0.5$. It is assumed that E_1 is given by the Rule of Mixtures, equation (4.12), and E_2, G_{12} and v_{12} are given by equations (4.23), (4.24) and (4.25). Values of v_f and v_m are given in Table 6.1. (Because E_y is complementary to E_x we do not need to plot it separately.)

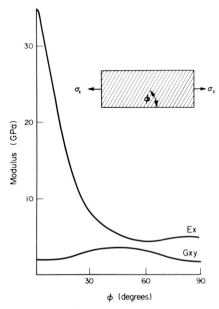

FIG. 4.13. Young's modulus and shear modulus oblique to fibre direction in a glass-epoxy lamina with $V_f = 0.5$. ($E_1 = 35$ GPa, $E_2 = 5$ GPa, $G_{12} = 1.9$ GPa and $v_{12} = 0.3$).

Figure 4.14 shows the interaction terms v_{xy} and η_{xyx} for the same composite. (η_{xyy} can be obtained from η_{xvx}.)

The shapes of the curves for the elastic constants depend a great deal on the degree of anisotropy of the lamina, the relative heights and sharpness of the peaks increasing with increasing anisotropy. For certain values of the constants E_x can exceed E_1 for a range of values of ϕ, while with other values E_x can be less than E_2. These situations are usually mutually exclusive; the former does not occur with glass-, carbon-, Kevlar-, or boron-polymers.

The existence of the shear stress–tensile stress interactions makes it difficult to perform off-axis tests on composites. For example, the grips normally used for tensile tests will not permit shear at the ends of the specimen. Consequently, during such tensile tests the specimens distort and buckle.

4.3.3. Oblique Strength Properties

Figure 4.15 shows the results of experiments to determine the effect of testing aligned

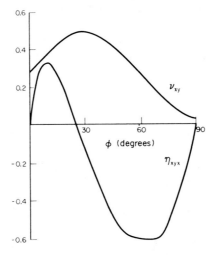

FIG. 4.14. Poisson's ratio, and interaction terms η_{xyx} oblique to fibre direction for the glass-epoxy lamina shown in Fig. 4.13.

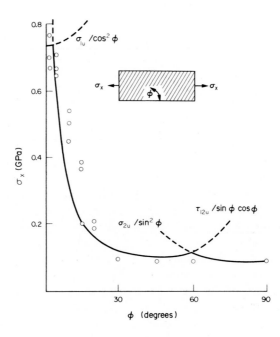

FIG. 4.15. Effect of fibre orientation on strength for silica–aluminium. Lines are drawn for maximum stress criterion for failure. (After Cooper, G. A. (1966) *J. Mech. Phys. Solids*, **14**, 103.)

silica–aluminium obliquely to the fibre direction. The strength falls off rapidly with angle.

These results can be explained quite well by a maximum stress theory. Let the laminate strengths in the principal directions be σ_{1u}, σ_{2u}, and τ_{12u}. For a uniaxial applied stress, σ_x, the stresses in the principal material directions can be obtained using the transformation matrix $[T]$ given in equation (4.32):

$$\begin{vmatrix} \sigma_1 \\ \sigma_2 \\ \tau_{12} \end{vmatrix} = [T] \cdot \begin{vmatrix} \sigma_x \\ \sigma_y \\ \tau_{xy} \end{vmatrix} \tag{4.43}$$

with $\sigma_y = \tau_{xy} = 0$. When σ_1 reaches σ_{1u}, σ_2 reaches σ_{2u}, and τ_{12} reaches τ_{12u}, equation (4.43) gives the following three limiting conditions:

$$\sigma_x = \sigma_{1u}/\cos^2 \phi \tag{4.44}$$

$$\sigma_x = \sigma_{2u}/\sin^2 \phi \tag{4.45}$$

$$\sigma_x = \tau_{12u}/\sin \phi \cos \phi \tag{4.46}$$

The ultimate strength in any direction ϕ to the fibres is the least of these values of σ_x for that value of ϕ. The curves shown in Fig. 4.15 were drawn using these equations. We observe three regions as ϕ increases:

1. For small ϕ, $\sigma_x = \sigma_{1u}/\cos^2 \phi$. This region ends when $\tan \phi = \tau_{12u}/\sigma_{1u}$, i.e. when σ_x given by equation (4.44) is equal to σ_x from equation (4.46).

2. For $\phi > \tan^{-1} (\tau_{12u}/\sigma_{1u})$ the strength is given by equation (4.46). This region ends when $\tan \phi = \sigma_{2u}/\tau_{12u}$.

3. For $\phi > \tan^{-1} (\sigma_{2u}/\tau_{12u})$ up to $\phi = \pi/2$, the strength is given by equation (4.45).

The maximum stress theory does not accurately predict the off-axis strength of reinforced polymer laminae. Figure 4.16 shows some results obtained with E-glass-epoxy. The curves with the cusps were drawn using equations (4.44) to (4.46). The experimental results fit a much smoother curve.

An alternative theory, which uses a yield criterion for anisotropic materials introduced by Hill, and involves six independent constants, provides the appropriate smooth curve. (It is often incorrectly referred to as a distortional energy failure theory.) When a lamina is under a general state of stress the Hill criterion reduces to

$$\frac{\sigma_1^2}{\sigma_{1u}^2} - \frac{\sigma_1 \sigma_2}{\sigma_{1u}^2} + \frac{\sigma_2^2}{\sigma_{2u}^2} + \frac{\tau_{12}^2}{\tau_{12u}^2} = 1. \tag{4.47}$$

For σ_{1u}, σ_{2u} and τ_{12u} we use failure rather than yield stresses, as suggested by Hill. Using equation (4.43), again with $\sigma_y = \tau_{xy} = 0$, and substituting into equation (4.46) values of σ_1, σ_2, and τ_{12} so obtained, we find

$$\frac{\cos^4 \phi}{\sigma_{1u}^2} + \left(\frac{1}{\tau_{12u}^2} - \frac{1}{\sigma_{1u}^2} \right) \cos^2 \phi \sin^2 \phi + \frac{\sin^4 \phi}{\sigma_{2u}^2} = \frac{1}{\sigma_x^2}. \tag{4.48}$$

The smooth curve shown in Fig. 4.16 was drawn using this equation.

(This approach has been developed further, by the introduction of more terms into the failure criterion. Although this improves the prediction of experimental results slightly, it is at the expense of much complication.)

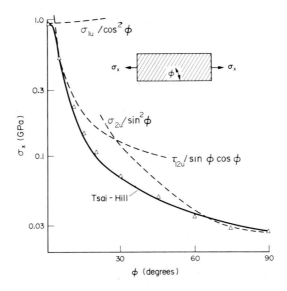

FIG. 4.16. Comparison of maximum stress criterion, and Tsai–Hill criterion for E-glass-epoxy, with experimental results. (After Tsai, S. (1968) *Fundamental Aspects of Fibre Reinforced Plastics Composites*, (ed. Schwartz, R. T. and Schwartz, H. S., *Interscience*, **3**.)

4.3.4. *Random Fibre Laminae*

Before we can calculate the moduli of random fibre laminae we need the appropriate stiffness matrix for aligned fibre laminates subject to off-axis strains:

$$\begin{vmatrix} \sigma_x \\ \sigma_y \\ \tau_{xy} \end{vmatrix} = \begin{vmatrix} \overline{Q}_{11} & \overline{Q}_{12} & \overline{Q}_{16} \\ \overline{Q}_{12} & \overline{Q}_{22} & \overline{Q}_{26} \\ \overline{Q}_{16} & \overline{Q}_{26} & \overline{Q}_{66} \end{vmatrix} \cdot \begin{vmatrix} \varepsilon_x \\ \varepsilon_y \\ \gamma_{xy} \end{vmatrix} \tag{4.49}$$

The $\overline{Q}_{ij}$ are determined by similar transformations to those used for the S_{ij}. The results also are very similar to those for the $\overline{S}_{ij}$ (equations (4.35)–(4.40)):

$$\overline{Q}_{11} = Q_{11}\cos^4\phi + 2(Q_{12} + 2Q_{66})\sin^2\phi\cos^2\phi + Q_{22}\sin^4\phi \tag{4.50}$$

$$\overline{Q}_{12} = (Q_{11} + Q_{22} - 4Q_{66})\sin^2\phi\cos^2\phi + Q_{12}(\sin^4\phi + \cos^4\phi) \tag{4.51}$$

$$\overline{Q}_{22} = Q_{11}\sin^4\phi + 2(Q_{12} + 2Q_{66})\sin^2\phi\cos^2\phi + Q_{22}\cos^4\phi \tag{4.52}$$

$$\overline{Q}_{16} = (Q_{11} - Q_{12} - 2Q_{66})\sin\phi\cos^3\phi + (Q_{12} - Q_{22} + 2Q_{66})\sin^3\phi\cos\phi \tag{4.53}$$

$$\overline{Q}_{26} = (Q_{11} - Q_{12} - 2Q_{66})\sin^3\phi\cos\phi + (Q_{12} - Q_{22} + 2Q_{66})\sin^3\phi\cos\phi \tag{4.54}$$

$$\overline{Q}_{66} = (Q_{11} + Q_{22} - 2Q_{12} - 2Q_{66})\sin^2\phi\cos^2\phi + Q_{66}(\sin^4\phi + \cos^4\phi). \tag{4.55}$$

We treat the lamina as made up of an infinite number of microlaminae with fibres in different directions. Each microlamina is strained by an amount ε_x. The in-plane modulus E_r is given by the total stress, divided by ε_x, i.e.

$$E_r = \frac{1}{\pi\varepsilon_x}\int_{-\frac{\pi}{2}}^{\frac{\pi}{2}}\sigma_x\,d\phi = \frac{1}{\pi}\int_{-\frac{\pi}{2}}^{\frac{\pi}{2}}\overline{Q}_{11}\,d\phi.$$

Performing the integration, remembering that $\cos^4 \phi = 1/8 \,(\cos 4\phi + 4\cos 2\phi + 3)$, etc., we find that

$$E_r = \tfrac{1}{8}(3Q_{11} + 2(Q_{12} + 2Q_{66}) + 3Q_{22}).$$

Substituting the engineering constants using equations (4.28) to (4.31) we obtain,

$$E_r = \frac{3E_1}{8(1 - v_{12}v_{21})} + \frac{v_{12}E_2}{4(1 - v_{12}v_{21})} + \frac{G_{12}}{2} + \frac{3E_2}{8(1 - v_{12}v_{21})}. \tag{4.56}$$

This simplifies, since $v_{12}v_{21} = v_{12}^2 E_2/E_1$, $v_{12} \simeq 0.3$ (equation (4.25)), and $E_2 \ll E_1$ for most aligned fibre laminates. Thus $1 - v_{12}v_{21} \simeq 1.00$. Also $G_{12} \simeq E_2/2.6$ (from equations (4.23) and (4.24) with $E_f \gg E_m$ and $v_m \simeq 0.3$, or with $v_f \simeq v_m \simeq 0.3$). Thus

$$E_r \simeq \tfrac{3}{8}E_1 + \tfrac{5}{8}E_2. \tag{4.57}$$

The strength is usually given by an analogous equation

$$\sigma_{ru} = \tfrac{3}{8}V_f\sigma_{fu} + V_m\sigma_{mu}. \tag{4.58}$$

We can also calculate the shear modulus. We carry out the same procedure as for tension, except that we apply a shear strain γ_{xy}. Thus we integrate $\bar{Q}_{66}$ with the result

$$G_r = \frac{G_{12}}{2} + \frac{E_1 + E_2(1 - 2v_{12})}{8(1 - v_{12}v_{21})}. \tag{4.59}$$

Making the same approximations as before, this reduces to

$$G_r \simeq \frac{E_1}{8} + \frac{E_2}{4}. \tag{4.60}$$

Thus G_r is approximately equal to $E_r/3$.

For composites having fibres which are random in three dimensions (these are not strictly laminae) the expressions usually used for modulus and strength are

$$E_r = \tfrac{1}{5}V_f E_f + V_m E_m \tag{4.61}$$

and

$$\sigma_r = \tfrac{1}{5}V_f\sigma_{fu} + V_m\sigma_{mu} \tag{4.62}$$

Equation (4.62) should be regarded as an upper bound, since the matrix may not be strained sufficiently, before the composite fractures, for it to be able to contribute its full strength.

In random fibre composites the fibres cannot be packed so closely as in aligned fibre laminae. Thus V_f cannot be so great (see e.g. Table 9.2). Because of this it is difficult to make the composite stronger and stiffer than the matrix when the fibres are random in three dimensions, where the packing problem is particularly severe. The maximum V_f achievable depends on fibre aspect ratio, being smaller for larger aspect ratios. However, if the fibres are more efficiently packed together by making them shorter, an aspect ratio penalty has to be taken into account. V_f in equation (4.62) must be multiplied by $(1 - s_c/2s)$, and in equation (4.61) by a factor (< 1) which depends on strain (see equation (4.19), and equation (4.21). For $s < s_c$ the strength penalty is even greater; equation (4.22)).

4.4. Laminates

Advanced composites are laminates of continuous aligned fibre composite laminae, or sometimes (for greater toughness) woven fibre composite laminae. The matrix is normally epoxy resin, but can be polyester for less exacting requirements, or polyimides, or other special polymers, when good properties at temperatures above about 150 C are required. Metal matrices are rarely used.

The laminates are designed to make the most efficient use of the directional nature of the laminae. Structural elements made from isotropic materials generally have redundant strength. The design objective for laminates is to arrange for the strength and stiffness in any direction to be no greater than that required by the expected loads, with the appropriate safety factors. Such a procedure should result in the minimum structural weight.

In this section a qualitative description of the main characteristics of laminates will be presented. For quantitative evaluations the student is referred to the reading list at the end of this chapter.

4.4.1. General Description of Laminates

The simplest types of fibre laminates are made by bonding together identical laminae, with particular fibre orientations with respect to each other. In cross-ply laminates the orientations are parallel to and at right angles to a particular direction. Usually fibres in adjacent layers are at right angles to each other. In angle-ply laminates each layer has the fibre oriented at either $+\phi$ or $-\phi$ to some particular direction (see Fig. 4.17).

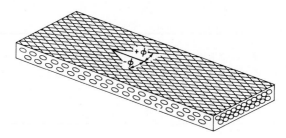

FIG. 4.17. Fibre directions in a two-layer laminate.

For special effects laminae may contain different types of fibre (hybrid composites) or have different matrices.

Laminates are often made so that they are symmetric about their central plane. A three-layer angle-ply laminate is symmetric if the two outer laminae are identical with respect to fibre direction and construction. If all three laminae have the same construction (including thickness, fibre volume fraction, etc.) then it is a regular symmetric angle-ply laminate.

An antisymmetric cross-ply or angle-ply laminate consists of an even number of layers, where adjacent layers have fibres at 90° to each other (cross-ply) or at some other fixed angle (angle-ply).

There is no universally agreed nomenclature for the orientation of the successive layers in laminates. They are usually indicated by a set of numbers together with + or − signs to indicate the layers. For example $[+30/-30/+45/-45/0/-45/+45/-30/+30]$ indicates a nine-layer laminate which is symmetric, and has layers with fibres oriented at $0°$, $\pm 30°$, and $\pm 45°$ to some particular direction. This particular notation is rather cumbersome, and more compact forms are being developed.

4.4.2. Laminate Properties

The mechanical properties of fibre laminates are quite different from isotropic homogeneous materials. This is because of the strongly directional nature of the mechanical properties of the layers, and because of the complex interactions between the layers.

Classical laminate theory dates from the late 1950s and can be expressed with apparent simplicity by matrix notation. The elastic properties are obtained by summing the transformed elastic constants $\overline{S}$ (equations 4.35–4.40) for each lamina divided by the lamina thickness, or $\overline{Q}_{ij}$ (equations 4.50–4.55) for each lamina multiplied by the lamina thickness. Thus for example, for n layers

$$\overline{Q}_{11} = \sum_{k=1}^{n} (\overline{Q}_{11k}t_k) \Big/ \sum_{k=1}^{n} t_k \tag{4.63}$$

where $\overline{Q}_{11k}$ is the transformed reduced stiffness (equation (4.50)) for the kth layer, which has thickness t_k.

However, the layers interact to produce extra effects, sometimes undesirable. A striking example of this is observed with a two-layer angle-ply laminate. When this laminate is stressed along an axis of symmetry (or any other axis) it twists (see Fig. 4.18) due to coupling between bending and extension. Symmetric laminates are usually used to avoid this problem.

The twisting effect due to coupling between bending and extension falls off with the number of layers, and becomes unimportant when there are more than 10 layers. It has its maximum effect when ϕ (Fig. 4.18) $= 45°$ i.e. a cross-ply laminate stressed along a bisector of the interply angle.

The same coupling results in an extension of the centre plane of the laminate when it is bent.

Sometimes coupling is required to produce special effects. For example, the twist in turbine blades can be produced by the appropriate laminate design, together with control of the residual stresses that arise during curing.

The analysis of laminate strength is more difficult, since the laminate does not necessarily fail when one lamina fails. A striking example of this is the cross-ply laminate, when stressed along one of the fibre directions. At relatively low stress, failure of the matrix occurs, in the laminae with the fibres at right angles to the stress (the matrix in these laminae suffers multiple cracking parallel to the fibres). The fibres are undamaged, and the composite can bear a great deal more load before the fibres start to fail. The result is a knee in the stress–strain curve, since the matrix failure causes a reduction in modulus. Angle-ply laminates do not behave like this; all laminae fail simultaneously, and no knee is observed.

Failure analysis requires iterative techniques, in which the load is increased in steps until it can be shown that all laminae have failed. For fibre-polymer laminates the most appropriate failure criterion is the Tsai–Hill yield failure criterion; equation (4.47).

Fig. 4.18. Twisting of an unbalanced laminate due to a stress along an axis of symmetry.

Excellent agreement between theory and experiment has been obtained using this for glass-epoxy laminates.

There are additional failure modes that do not occur at all with isotropic materials. For example, the laminate can split apart due to interlaminar stresses. These stresses are very high at the edge of the laminate, but at about one laminate thickness distant from the edge they become negligible. They are present near any free edge, such as a round hole. However, the tensile stress is affected by the stacking sequence of the laminae, and splitting can often be prevented by using the appropriate sequence.

Further Reading

JONES, R. M., (1975) *Mechanics of Composite Materials* (McGraw-Hill, New York).
ASHTON, J. E., HALPIN, J. C. and PETIT, P. H., (1969) *Primer on Composite Materials; Analysis* (Technomic Publishing Co., Stamford, Conn.).

5

Reinforcement Processes

AN UNDERSTANDING of reinforcement processes is important for the production of high quality composites. In addition, knowledge of these processes, together with failure mechanisms described in the next two chapters, should promote the development of composites designed for optimum toughness and strength.

The interactions between fibres and matrix are extremely complex, and imperfectly understood. The first attempt to explain the reinforcing effect of the fibres was based entirely on elastic interactions. This was first described by Cox in 1952, and is now referred to as the shear lag theory. These ideas have been refined by others, but basically only explain the behaviour of composites at low stress.

This was followed in 1964 by slip theories, based on matrix plasticity at the fibre surface near the fibre ends and frictional sliding of the fibre ends. (Cottrell, Kelly, and Outwater.) Matrix plasticity is applicable to well-bonded reinforced metals, and frictional sliding to reinforced polymers and ceramics.

Recent work has produced a set of expressions that can be used to describe slip as well as elastic stress transfer; in this chapter these expressions are developed and discussed. The reader should fully understand Section 4.1, which presents many of the basic ideas used here (and in Chapter 6) before attempting this chapter.

A selection of the important early contributions and reviews appears in the reading list at the end of this chapter. A complete list of the nomenclature used appears in the Appendix.

5.1. Elastic Stress Transfer

We assume that both fibres and matrix behave elastically, and that the interface transfers the stress from fibres to matrix without yielding or slip. Each fibre is surrounded by other fibres which are packed in an orderly fashion, for example, hexagonally, as shown in Fig. 5.1. The matrix tensile strain is assumed to be constant at the ring of nearest neighbours, at distance R from the fibre we are considering, and we assume that we can equate this matrix tensile strain, ε_m, with the composite tensile strain, ε_l.

We thus have a stress distribution with circular symmetry, i.e. the matrix displacement, w (Fig. 5.1), and the shear stress, τ, do not vary with orientation about the fibre axis. They are both, however, a function of the radial distance, z, from the fibre centre.

Equate the shear forces at distance z with those at the fibre surface ($z = r$) in the composite element shown in Fig. 5.1:

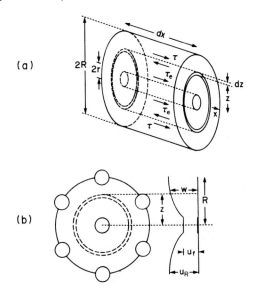

FIG. 5.1. (a) Short length of fibre and surrounding matrix, (b) fibre with nearest neighbours, hexagonally packed, and associated displacements along the fibre axis.

$$2\pi z \tau dx = 2\pi r \tau_e dx$$

or

$$\tau = r\tau_e/z.$$

Now the ratio of τ, to the shear strain, dw/dz, is equal to the matrix shear modulus G_m, so that

$$\frac{dw}{dz} = \frac{\tau}{G_m} = \frac{\tau_e r}{G_m z}.$$

This equation is integrated:

$$\int_{u_f}^{u_R} dw = \frac{\tau_e r}{G_m} \int_r^R \frac{dz}{z}$$

to give,

$$u_R - u_f = \frac{\tau_e r}{G_m} \ln (R/r). \qquad (5.1)$$

The value of R/r depends on the fibre packing. Consider square packing, Fig. 5.2. The box ABCD has area R^2, and includes one complete fibre cross-section (the sum of the quarter sections at each of the four corners). Thus the volume fraction of the fibres, V_f, which is equal to the area fraction, is $V_f = \pi r^2/R^2$, so that

$$\ln (R/r) = \tfrac{1}{2} \ln (\pi/V_f). \qquad (5.2)$$

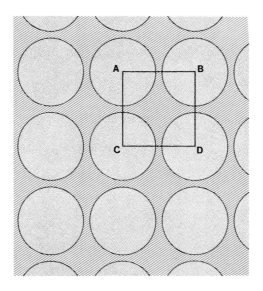

FIG. 5.2. Fibres in a square-packed arrangement.

Similarly, for hexagonal packing

$$\ln (R/r) = \tfrac{1}{2} \ln (2\pi/\sqrt{3} V_f).$$

(5.3)

(The reader should derive this as an exercise). The difference between the two results is quite small. For $V_f = 0.50$, for example, square packing gives 0.919 for $\ln (R/r)$ while hexagonal packing gives 0.991. We will write

$$\ln (R/r) = \tfrac{1}{2} \ln (P_f/V_f)$$

(5.4)

where P_f is the packing factor.

Substituting equation (5.4) into equation (5.1) and rearranging gives

$$\tau_e = \frac{E_m(u_R - u_f)}{(1 + v_m)r \ln (P_f/V_f)}$$

(5.5)

(remember that $G = E/2(1 + v)$).

τ_e transfers stress to the fibres, as described in Section 4.1. Equation (4.1) can be rewritten

$$\frac{d\sigma_f}{dx} = -\frac{2\tau_e}{r}$$

(5.6)

and substituting for τ_e from equation (5.5),

$$\frac{d\sigma_f}{dx} = -\frac{2E_m(u_R - u_f)}{(1 + v_m)r^2 \ln (P_f/V_f)}.$$

(5.7)

The fibre displacement, u_f, can be calculated from the fibre stress, since the fibre strain is

$\varepsilon_f = du_f/dx$. Thus

$$\frac{du_f}{dx} = \sigma_f/E_f.$$

At $z = R$, $du_R/dx = \varepsilon_m$ which we assume $= \varepsilon_1$. We can, therefore, differentiate equation (5.7) and substitute for du_R/dx and du_f/dx:

$$\frac{d^2\sigma_f}{dx^2} = \frac{2E_m(\varepsilon_1 - \sigma_f/E_f)}{(1 + v_m)r^2 \ln(P_f/V_f)}.$$

We now introduce the dimensionless parameter, n, where

$$n^2 = 2E_m/\{E_f(1 + v_m) \ln(P_f/V_f)\} \tag{5.8}$$

and the differential equation may be written

$$\frac{d^2\sigma_f}{dx^2} = \frac{n^2}{r^2}(\sigma_f - E_f\varepsilon_1). \tag{5.9}$$

This has the solution

$$\sigma_f = E_f\varepsilon_1 + B \sinh(nx/r) + D \cosh(nx/r) \tag{5.10}$$

where B and D are constants determined by the boundary conditions. Assume that no stress is transferred across the fibre ends, i.e. $\sigma_f = 0$ at $x = L$ and $x = -L$ (see Fig. 5.3). Thus $B = 0$ and $D = -E_f\varepsilon_1/\cosh(ns)$, where $s = L/r$, the fibre aspect ratio. Equation (5.10) becomes

$$\sigma_f = E_f\varepsilon_1\{1 - \cosh(nx/r)/\cosh(ns)\}. \tag{5.11}$$

Differentiation of this, and multiplication of the result by $r/2$ (see equation (5.6)) gives us τ_e:

$$\tau_e = \tfrac{1}{2}nE_f\varepsilon_1 \sinh(nx/r)/\cosh(ns). \tag{5.12}$$

The fibre stress and surface shears are shown schematically in Fig. 5.3. This should be contrasted with Fig. 4.1 which shows stresses and displacements due to slip.

To determine the composite stress we use the Rule of Averages, equation (4.5), so we must first calculate the average fibre stress, $\bar{\sigma}_f$. We integrate equation (5.11) for this;

$$\bar{\sigma}_f = \frac{E_f\varepsilon_1}{L} \int_0^L \{1 - \cosh(nx/r)/\cosh(ns)\}\, dx$$

$$= E_f\varepsilon_1(1 - \tanh(ns)/ns)$$

and for the matrix assume that $\bar{\sigma}_m = E_m\varepsilon_1$. Substituting $\bar{\sigma}_f$ and $\bar{\sigma}_m$ into equation (4.5) gives

$$\sigma_1 = \{V_f E_f(1 - \tanh(ns)/ns) + V_m E_m\}\varepsilon_1. \tag{5.13}$$

This is the stress–strain relationship for the composite.

The maximum fibre stress, $\sigma_{f\text{max}}$, is at the centre, and is given by equation (5.11) with $x = 0$:

$$\sigma_{f\text{max}} = E_f\varepsilon_1\{1 - \text{sech}(ns)\}. \tag{5.14}$$

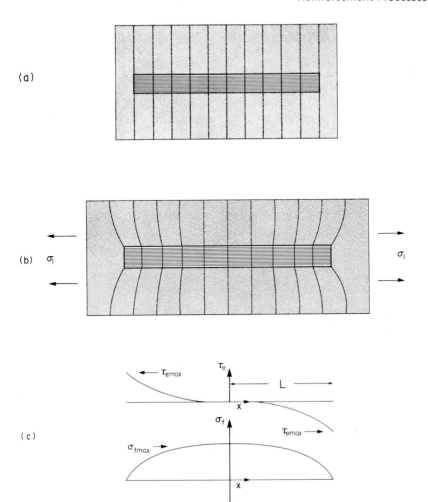

FIG. 5.3. Single fibre composite element: (a) unstressed and (b) stressed. (c) shows the fibre–matrix interfacial stress and the fibre internal stress for elastic stress transfer.

This theory is approximate. We have neglected the stress transfer across the fibre ends, and the stress concentration there, which greatly increases τ_e at the ends. We have also neglected the extra load thrown on the ring of fibres surrounding the end region of each fibre, and the extra load on the matrix in this region. These effects will all be unimportant for $s > 10$, except for the underestimation of τ_e. The increased τ_e at the fibre ends will affect the onset of the slip processes discussed later.

5.2. Elastic Stress–Strain Relationships

Equation (5.13) states that stress is directly proportional to strain. Thus all stress–strain curves are straight lines in the region governed by this equation.

For *ns* values greater than 3.0, tanh $(ns) = 1.00$ with an error of less than 0.5%. So for very long fibres $(s \to \infty)$ equation (5.13) becomes

$$\sigma_1 = (V_f E_f + V_m E_m)\varepsilon_1 \tag{5.15}$$

and the composite modulus, $E_1 = \sigma_1/\varepsilon_1$, obeys the Rule of Mixtures (equation (4.12)).

Short-fibre composites will deviate from the mixture rule by an amount depending on *n* as well as *s*. Table 5.1 lists some *n* values for $V_f = 0.5$ for hexagonal packing $(P_f = 2\pi/\sqrt{3}$; the corresponding values for square packing, $P_f = \pi$, may be obtained by multiplying these values by 1.04).

A deviation from the mixture rule by 1% or more occurs when $ns < 100$. Since most of the metal–fibre combinations have $n > 0.2$, significant deviations only occur with aspect ratios less than 500. These are quite short fibres; carbon fibres with an aspect ratio of 500 are only about 4 mm long.

Reinforced polymers require longer fibres, since *n* is smaller. Stress–strain curves are shown in Fig. 5.4 for E-glass-epoxy. Note the large deviation when $s = 10$. The deviation for $s = 500$ is only about 1.2%.

The curves shown in Fig. 5.4 have been terminated at the highest stress that can be applied before slip occurs at the fibre–matrix interface. The lines are drawn on the assumption that the adhesion is as good as it theoretically can be, i.e. that slip starts when the highest interfacial shear stress (neglecting the stress concentrations at the fibre ends) reaches the matrix shear yield strength.

The maximum value of τ_e is at the fibre end, $x = L$, so from equation (5.12) we find

$$\tau_{emax} = \tfrac{1}{2} n E_f \varepsilon_1 \tanh (ns). \tag{5.16}$$

Let the strain at the onset of slip be ε_{1s}. Writing τ_{my} for the matrix shear yield strength, we can rewrite equation (5.16) for the onset of slip $(\tau_{emax} = \tau_{my})$ as

$$\varepsilon_{1s} = 2\tau_{my} \coth (ns)/nE_f. \tag{5.17}$$

TABLE 5.1. *Values of n for Various Fibre–Matrix Combinations for* $V_f = 0.5$ *and* $P_f = 2\pi/\sqrt{3}$

Fibre	E_f (GPa)	Matrix (E_m, GPa)			
		Epoxy (2.5)	Polyimide (3.1)	Aluminium (71)	Nickel (200)
Nylon	2.9	0.81	–	–	–
Wool	6	0.56	0.63	–	–
E-glass	72	0.162	0.181	0.86	–
Kevlar	130	0.120	0.135	–	–
Asbestos[†]	160	0.108	0.121	0.58	–
Steel	210	0.095	0.110	0.50	0.86
Beryllium	315	0.078	0.086	0.41	0.70
Tungsten	350	0.074	0.082	0.39	0.66
Carbon[‡]	377	0.071	0.079	0.38	0.64
Boron	420	0.067	0.075	0.36	0.61
SiC	480	0.063	0.070	0.33	0.57
SiC[§]	840	0.048	0.053	0.252	0.43
Al_2O_3[§]	2250	0.029	0.032	0.154	0.262

[†] Chrysotile.
[‡] Stiff carbon.
[§] Whiskers, grown for maximum Young's modulus.

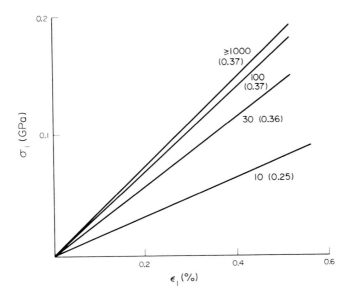

FIG. 5.4. Theoretical stress–strain curves for E-glass-epoxy up to the slip point for aspect ratios 10, 30, 100 and > 1000, as marked. The maximum fibre stresses (GPa) at the slip point are also indicated.

The corresponding composite stress, σ_{1s} is obtained by substituting ε_{1s} into equation (5.13):

$$\sigma_{1s} = \frac{2\tau_{my}}{nE_f}\left\{(V_fE_f + V_mE_m)\coth(ns) - \frac{V_fE_f}{ns}\right\} \qquad (5.18)$$

Equations (5.17) and (5.18) define the SLIP POINT for the case of perfect adhesion. We can evaluate the corresponding maximum fibre stress by substituting ε_{1s} into equation (5.14):

$$\sigma_{fmax} = 2\tau_{my}\tanh(ns/2)/n. \qquad (5.19)$$

The values of σ_{fmax} have been indicated on Fig. 5.4. These should be contrasted with the ultimate fibre strength of 3.4 GPa (Table 3.2). It is clear that slip occurs when the fibre is bearing little more than one-tenth of its breaking-stress, even when the fibres are very long. (For $s \to \infty$ the maximum fibres is 0.375 GPa). In practice slip will start even sooner than this, since τ_{emax} is greater than indicated by equation 5.16, due to the stress concentration at the fibre end. If the adhesion is imperfect the stress and strain at the slip point will be even less.

5.3. Reinforced Metals

Slip between fibres and matrix will normally take place at an interfacial shear stress which varies along the fibre length. However, in the case of metal matrices, if we neglect work hardening, and if the fibre–matrix adhesion is perfect, the shear stress will be constant in the slip regions shown in Fig. 5.5, and equal to the shear yield stress of the

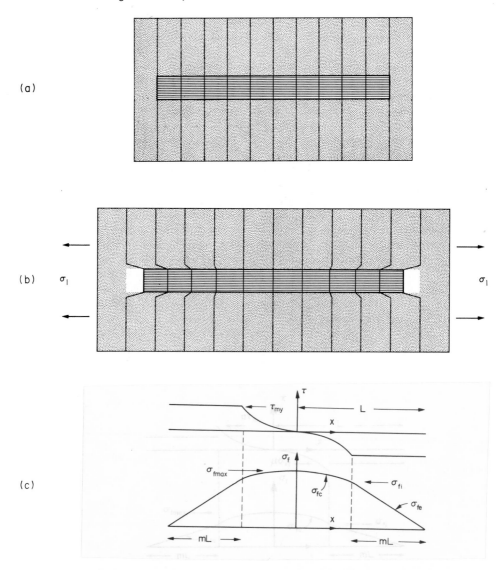

FIG. 5.5. Single fibre composite element: (a) unstressed and (b) stressed. (c) shows the fibre–matrix interfacial stress and the fibre internal stress for a well-bonded reinforced metal.

metal, τ_{my}. The slip occurs over a length mL at both ends of the fibre, where m is a dimensionless parameter that depends on the applied stress. Near the fibre mid-section the interfacial shears are generated by elastic interactions.

The composite obeys the Rule of Averages, equation (4.5). At this stage we assume that $\bar{\sigma}_m = E_m\varepsilon_1$, so the problem reduces to determining $\bar{\sigma}_f$. This is done by treating the fibre end and centre regions separately.

For the centre region we use the elasticity treatment (Section 5.1) up to equation (5.10). To determine the constants B and D in this equation we use the boundary conditions $\sigma_f = \sigma_{fi}$ at $x = \pm L(1 - m)$, as shown in Fig. 5.5. σ_{fi} is the stress transferred to the fibre in the slip region, and will be determined later. Insertion of these boundary conditions into equation (5.10) gives

$$\sigma_{fc} = E_f \varepsilon_1 + (\sigma_{fi} - E_f \varepsilon_1) \cosh(nx/r)/\cosh(n\bar{s}) \qquad (5.20)$$

where we have used σ_{fc} for the fibre stress in the centre region, and $\bar{s}$ for the reduced aspect ratio:

$$\bar{s} = s(1 - m). \qquad (5.21)$$

The elastic contribution to the fibre stress is determined by the magnitude of the maximum interfacial shear stress. The highest shears are at the boundaries of the elastic stress transfer region ($x = \pm L(1 - m)$) and are equal to τ_{my}. We proceed as follows. Differentiate equation 5.20:

$$\frac{d\sigma_{fc}}{dx} = \frac{n}{r}(\sigma_{fi} - E_f \varepsilon_1) \sinh(nx/r)/\cosh(n\bar{s}).$$

Evaluate the derivative at $x = L(1 - m)$ and equate with $-2\tau_{my}/r$, since the elastic region is governed by equation (5.6). Thus,

$$-\frac{2\tau_{my}}{r} = \frac{n}{r}(\sigma_{fi} - E_f \varepsilon_1) \tanh(n\bar{s}).$$

This can be rearranged to give

$$\sigma_{fi} - E_f \varepsilon_1 = -\frac{2\tau_{my}}{n} \coth(n\bar{s}). \qquad (5.22)$$

and substituting this into equation (5.20)

$$\sigma_{fc} = E_f \varepsilon_1 - \frac{2\tau_{my}}{n} \cosh(nx/r)/\sinh(n\bar{s}). \qquad (5.23)$$

The average stress in this region is

$$\bar{\sigma}_{fc} = \frac{1}{L(1 - m)} \int_0^{L(1 - m)} \sigma_{fc} dx$$

so that on substituting for σ_{fc} and carrying out the integration we obtain

$$\bar{\sigma}_{fc} = E_f \varepsilon_1 - \frac{2\tau_{my}}{n^2 s(1 - m)}. \qquad (5.24)$$

The maximum fibre stress is the value of σ_{fc} at $x = 0$. Using equation (5.23) we find

$$\sigma_{f\max} = E_f \varepsilon_1 - \frac{2\tau_{my}}{n} \operatorname{cosech}(n\bar{s}). \qquad (5.25)$$

We now turn to the fibre ends. Here the interfacial shear stress is constant at $\tau_i = \tau_{my}$. Thus for equation (4.1) we have

$$\frac{d\sigma_{fe}}{dx} = -\frac{2\tau_{my}}{r} \qquad (5.26)$$

where σ_{fe} denotes the fibre stress in the slip region. Integrating equation (5.26) with the boundary condition $\sigma_{fe} = 0$ at $x = L$ we find

$$\sigma_{fe} = \frac{2\tau_{my}}{r}(L - x) \tag{5.27}$$

and since $\sigma_{fe} = \sigma_{fi}$ at $x = L(1 - m)$ (see Fig. 5.5)

$$\sigma_{fi} = 2ms\tau_{my}. \tag{5.28}$$

The average fibre stress in the slip region is obtained by integration:

$$\bar{\sigma}_{fe} = \frac{1}{mL} \int_{L(1-m)}^{L} \sigma_{fe} \, dx \tag{5.29}$$

and substituting for σ_{fe} from equation (5.27) and integrating gives

$$\bar{\sigma}_{fe} = ms\tau_{my}. \tag{5.30}$$

The average stress for the whole fibre is found by adding $\bar{\sigma}_{fc}$ and $\bar{\sigma}_{fe}$ in the correct proportions:

$$\bar{\sigma}_f = (1 - m)\bar{\sigma}_{fc} + m\bar{\sigma}_{fe} \tag{5.31}$$

substituting for $\bar{\sigma}_{fc}$ and $\bar{\sigma}_{fe}$ using equations (5.24) and (5.30) gives

$$\bar{\sigma}_f = E_f\varepsilon_1(1 - m) - \frac{2\tau_{my}}{n^2 s} + m^2 s\tau_{my} \tag{5.32}$$

m still has to be determined. This can be done by eliminating σ_{fi} between equations (5.28) and (5.22). Thus

$$m = \frac{E_f\varepsilon_1}{2\tau_{my}s} - \frac{\coth(n\bar{s})}{ns}. \tag{5.33}$$

This equation requires numerical methods for its solution. However, if we consider, for the moment, only fibres with aspect ratios of at least twice the critical, we find that m cannot exceed 0.5 (compare equations (4.18) and (5.28), remembering that σ_{fi} cannot exceed σ_{fu}, the ultimate fibre strength) and since s_c is usually greater than 10, $\coth(n\bar{s}) \simeq 1.00$ with better than 1% accuracy, ($s_c = \sigma_{fu}/2\tau_{my}$ in this case, see equation (4.14)).

Making this approximation, equation (5.33) becomes

$$m \simeq \frac{E_f\varepsilon_1}{2\tau_{my}s} - \frac{1}{ns} \tag{5.34}$$

and substituting equation (5.34) into equation (5.32) gives

$$\bar{\sigma}_f \simeq E_f\varepsilon_1 - \frac{\tau_{my}}{n^2 s} - \frac{E_f^2\varepsilon_1^2}{4\tau_{my}s}. \tag{5.35}$$

Substitution of $\bar{\sigma}_f$ into the Rule of Averages, equation (4.5), with $\bar{\sigma}_m = E_m\varepsilon_1$ gives the stress–strain relationship

$$\sigma_1 \simeq (V_f E_f + V_m E_m)\varepsilon_1 - \frac{V_f}{s}\left(\frac{\tau_{my}}{n^2} + \frac{E_f^2\varepsilon_1^2}{4\tau_{my}}\right). \tag{5.36}$$

This is the same as equation (4.9) except for the additional term $-V_f\tau_{my}/n^2 s$, which arises from the elastic interaction. (Note that in this case $\tau_i = \tau_{my}$). Equation (5.36) reduces to the mixture rule expression, equation (4.12), when $s \to \infty$.

The stress–strain curve now has two parts. The first part is governed by elastic stress transfer, equation (5.13), given with sufficient accuracy by the approximate form:

$$\sigma_1 \simeq (V_f E_f + V_m E_m)\varepsilon_1 - \frac{V_f E_f \varepsilon_1}{ns}. \tag{5.37}$$

This region ends at the slip point, when $\varepsilon_1 = \varepsilon_{1s}$, given with sufficient accuracy by

$$\varepsilon_{1s} \simeq 2\tau_{my}/nE_f \tag{5.38}$$

from equation (5.17). Note that we have replaced τ_{mu} by τ_{my}. These are equal for non-work hardening metals.

Yielding and slip take place when $\varepsilon_1 > \varepsilon_{1s}$ and the stress–strain trajectory is given by equation 5.36. This is noticeably curved for aspect ratios close to $2s_c$.

Figure 5.6 shows stress–strain curves for boron–magnesium alloy derived using these equations. (This material was chosen because the fibre breaking-strain (0.83 %) is about the same as the matrix yield strain, so that we need not be concerned with failure mechanisms.) It will be seen that the stress–strain trajectory for $s = 20$ is noticeably curved. ($s_c = 10$). The dashed lines are the elastic stress–strain trajectories, and for $s = 20$ the elastic line has been extrapolated to ε_{fu}, to emphasize the effect of slip.

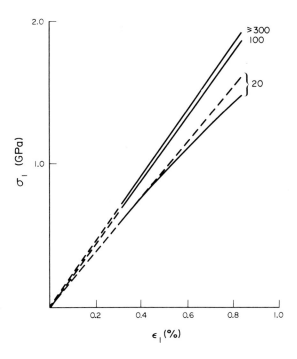

FIG. 5.6. Theoretical stress–strain curves for B–Mg alloy. The curves have been terminated at ε_{fu}. Aspect ratios of 30, 100, and > 300 as marked. $V_f = 0.5$. Dotted lines are elastic trajectories.

In the case of poorly adhering fibres the governing equations are those used for reinforced ceramics (see Section 5.5).

5.4. Reinforced Polymers

Frictional slip usually takes place between fibres and polymers, instead of the matrix flow that occurs at the interface in well-bonded reinforced metals. In addition, we must consider the effect of poor interfacial adhesion, since this is common with inert fibres such as carbon.

First consider the effect of poor adhesion. This will affect the onset of slip, i.e. the slip point. When adhesion is perfect, the strain at the slip point is ε_{1s} (equation 5.17). When it is imperfect, we introduce a non-dimensional parameter, a, which cannot exceed unity. At the slip point we make τ_{emax} (equation (5.16)) equal to $a\tau_{my}$ rather than τ_{my}. ε_{1s} is, therefore, reduced:

$$\varepsilon_{1s} = 2a\tau_{my} \coth{(ns)}/nE_f \tag{5.39}$$

and so is σ_{1s}:

$$\sigma_{1s} = \frac{2a\tau_{my}}{nE_f} \left\{ (V_f E_f + V_m E_m) \coth{(ns)} - \frac{V_f E_f}{ns} \right\}. \tag{5.40}$$

The derivative of σ_{fc} at $x = L(1-m)$ is also reduced by poor adhesion. Thus,

$$\frac{d\sigma_{fc}}{dx} = -\frac{2a\tau_{my}}{r}$$

at this point, and equation (5.22) becomes

$$\sigma_{fi} - E_f \varepsilon_1 = -\frac{2a\tau_{my}}{n} \coth{(ns)}. \tag{5.41}$$

Similarly, equation (5.24) becomes

$$\bar{\sigma}_{fc} = E_f \varepsilon_1 - \frac{2a\tau_{my}}{n^2 s(1-m)}. \tag{5.42}$$

The effect of these changes is that τ_{my} is replaced in the equations with $a\tau_{my}$, and the elastic stress transfer is thereby reduced.

For very poor adhesion, e.g. $a = 10^{-3}$, the stress and strain at the slip point are extremely small. In such cases elasticity effects can be neglected, and the whole stress–strain curve computed on the basis of slip effects.

The interfacial friction stress is the product of the coefficient of friction, μ, and the compressive stress normal to the fibre–matrix interface, σ_{rt}. For reinforced polymers we need consider only two contributors to σ_{rt}:

(1) the residual stress which results from differences between fibre and matrix contractions during manufacture, and

(2) the stress due to Poisson's contractions of the matrix when a stress is applied to the composite.

The residual stress, σ_r, is a constant which has seldom been measured, but can be estimated from the known contractions during manufacture, together with a knowledge of the temperature at which the matrix creep rate becomes negligible.

The matrix Poisson's contraction can be determined, in principle, by an elasticity analysis. However, apart from the unrealistic case of a single fibre in a tube of matrix, no suitable expressions have yet been developed for it. The stress is governed by the matrix and composite tensile strains, so we can express it by the general form $-v_1 E_m \varepsilon_1$. v_1 is the effective Poisson's ratio, which is some function of the elastic constants and volume fractions of fibres and matrix. It is unlikely to be very different from v_m.

The expression for σ_{rt} is, therefore,

$$\sigma_{rt} = \sigma_r - v_1 E_m \varepsilon_1. \tag{5.43}$$

The frictional shear stress is $-\mu \sigma_{rt}$, and only exists if σ_{rt} is less than zero (i.e. compressive).

The treatment now follows that for reinforced metals, except that we replace τ_{my} by $-\mu \sigma_{rt}$ in equation (5.26):

$$\frac{d\sigma_{fe}}{dx} = \frac{2\mu}{r}(\sigma_r - v_1 E_m \varepsilon_1) \tag{5.44}$$

and this on integration, with $\sigma_{fe} = 0$ at $x = L$, gives

$$\sigma_{fe} = \frac{2\mu}{r}(v_1 E_m \varepsilon_1 - \sigma_r)(L - x) \tag{5.45}$$

so that instead of equation (5.30) for $\bar{\sigma}_{fe}$ we have

$$\bar{\sigma}_{fe} = \mu m s (v_1 E_m \varepsilon_1 - \sigma_r) \tag{5.46}$$

and consequently our average fibre stress, $\bar{\sigma}_f$, (equation (5.31)) comes to

$$\bar{\sigma}_f = E_f \varepsilon_1 (1 - m) - \frac{2a\tau_{my}}{n^2 s} + \mu m^2 s (v_1 E_m \varepsilon_1 - \sigma_r) \tag{5.47}$$

using equation (5.42) for $\bar{\sigma}_{fc}$ and equation (5.46) for $\bar{\sigma}_{fe}$. For σ_{fi} instead of equation (5.28) we have

$$\sigma_{fi} = 2\mu m s (v_1 E_m \varepsilon_1 - \sigma_r) \tag{5.48}$$

m is obtained by elimination of σ_{fi} between equations (5.48) and (5.41):

$$m = \frac{E_f \varepsilon_1 - 2a\tau_{my} \coth(n\bar{s})/n}{2\mu s (v_1 E_m \varepsilon_1 - \sigma_r)}. \tag{5.49}$$

To avoid having to use numerical methods to determine m, we consider fibres sufficiently long that $\coth(n\bar{s}) \simeq 1.00$. For this we define the critical aspect ratio by the equation:

$$s_c = -\sigma_{fu}/2\mu\sigma_r. \tag{5.50}$$

This is equivalent to s_c in equation (4.14), except for the exclusion of the strain-dependent term in τ_i, i.e. $v_1 E_m \varepsilon_1$. The fibres must have $s \geqslant 2s_c$, so that we can write

$$m \simeq \frac{E_f \varepsilon_1 - 2a\tau_{my}/n}{2\mu s (v_1 E_m \varepsilon_1 - \sigma_r)}. \tag{5.51}$$

For our stress–strain relation we use the Rule of Averages (equation (4.5)), together with equation (5.47) for $\bar{\sigma}_f$, equation (5.51) for m, and with $\sigma_m = E_m\varepsilon_1$:

$$\sigma_1 = \{V_f E_f + V_m E_m\}\varepsilon_1 - \frac{V_f}{s}\left\{\frac{(E_f\varepsilon_1)^2 - (2a\tau_{my}/n)^2}{4\mu(v_1 E_m\varepsilon_1 - \sigma_r)} + \frac{2a\tau_{my}}{n^2}\right\}. \tag{5.52}$$

Figure 5.7 is a schematic drawing of the fibre tensile and interfacial shear stresses. Note the discontinuity in the interfacial stress at $x = L(1 - m)$ which arises because the adhesion strength is different from the friction stress.

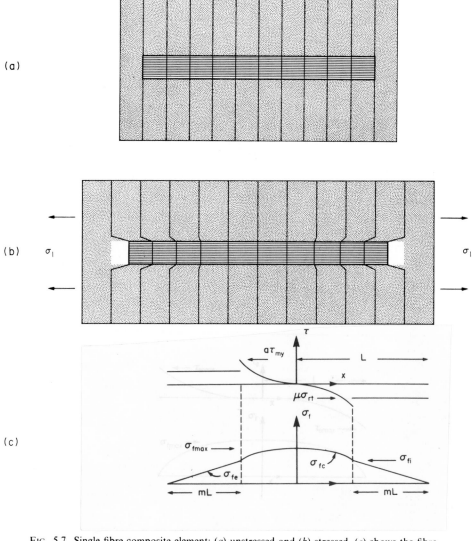

FIG. 5.7. Single fibre composite element: (*a*) unstressed and (*b*) stressed. (*c*) shows the fibre–matrix interfacial stress and the fibre internal stress for imperfectly bonded reinforced polymers ($a < 1$).

Fig. 5.8 shows the stress–strain curves given by this equation for carbon-epoxy, with $V_f = 0.5$, so that $n = 0.0707$ (Table 5.1). The other values used were $\sigma_r = -30$ MPa, $\mu = 0.2$, $v_1 = 0.34$ ($= v_m$) and $a = 1.00$ (perfect adhesion). With these values the slip point occurs at $\varepsilon_{1s} = 0.23\%$, and the minimum aspect ratio for which these equations can be used is $2s_c = 350$. Some curvature in the stress–strain trajectory is noticeable at the lowest aspect ratio. Notice that s_c is more than an order of magnitude greater than for the boron–magnesium alloy. The stress–strain curves have been terminated at the failure strain of the fibres.

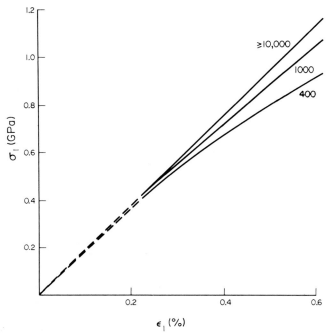

FIG. 5.8. Theoretical stress–strain curves for carbon-epoxy, with aspect ratios indicated on the curves; $V_f = 0.5$. The dashed lines indicate the elastic region.

5.5. Reinforced Ceramics and Cements

With reinforced Portland cements and similar materials (e.g. plasters) the equations used for reinforced polymers are appropriate, and since the ultimate tensile strain of the cement is very small the matrix Poissons shrinkage term may be neglected. Thus, beyond the slip point,

$$\sigma_1 = \{V_f E_f + V_m E_m\}\varepsilon_1 - \frac{V_f}{s}\left\{\frac{(E_f \varepsilon_1)^2 - (2a\tau_{mu}/n)^2}{-4\mu\sigma_r} + \frac{2a\tau_{mu}}{n^2}\right\}. \qquad (5.53)$$

(Note that we use τ_{mu} instead of τ_{my}.) s_c is given by equation (5.50). Below the slip point we use equation (5.13).

In the case of reinforced ceramics, the best systems are those for which σ_r is very small. Thus we cannot neglect matrix Poisson's shrinkage, in this case. In addition, the moduli of fibres and matrix are not very different, and a fibre Poisson's shrinkage term must also be included in the equations. The fibre contraction stress has not been satisfactorily evaluated. However, the fibre strain due to its Poisson's contraction is $-v_f\sigma_f/E_f$. If this is translated directly into a matrix stress, the matrix would be in tension with a stress of $v_f\sigma_f E_m/E_f$. We will express this stress more generally by $v_2\sigma_f E_m/E_f$, where v_2 is a function of the elastic constants and the volume fraction, as yet undetermined. However, it is unlikely that v_2 and v_f are very different.

The analysis for reinforced ceramics follows that for polymers up to equation (5.43), except that we replace τ_{my} by τ_{mu}. Equation (5.43) becomes

$$\sigma_{rt} = \sigma_r - v_1 E_m \varepsilon_1 + v_2 \sigma_{fe} E_m/E_f. \tag{5.54}$$

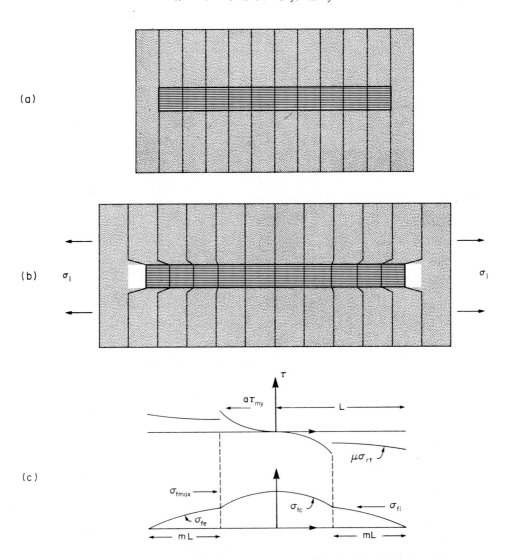

FIG. 5.9. Single fibre composite element: (a) unstressed and (b) stressed. (c) shows the fibre–matrix interfacial stress and the fibre internal stress for imperfectly bonded reinforced ceramics ($a < 1$).

As long as $\sigma_{rt} < 0$ we can determine σ_{fe} by integration, as before,

$$\sigma_{fe} = E_f(v_1 E_m \varepsilon_1 - \sigma_r)\{1 - exp(-2\mu v_2 E_m(L-x)/E_f r)\}/E_m v_2 \qquad (5.55)$$

The interfacial shear stress near the fibre ends (see e.g. equation (4.1) is proportional to $d\sigma_{fe}/dx$, and hence is not constant. Instead, it increases towards the fibre ends, as shown schematically in Fig. 5.9.

We determine $\bar{\sigma}_{fe}$ by integration (see equation (5.29)). Write

$$p = 2\mu v_2 ms E_m/E_f \qquad (5.56)$$

then

$$\bar{\sigma}_{fe} = E_f(v_1 E_m \varepsilon_1 - \sigma_r)\{1 - (1 - e^{-p})/p\}/E_m v_2 \qquad (5.57)$$

and equation (5.55) at $x = L(1 - m)$ gives

$$\sigma_{fi} = E_f(v_1 E_m \varepsilon_1 - \sigma_r)(1 - e^{-p})/E_m v_2. \qquad (5.58)$$

The expression for m is obtained by eliminating σ_{fi} between equations (5.58) and (5.41):

$$m = -\frac{E_f}{2\mu v_2 s E_m} \ln\left\{1 - \frac{E_m v_2[nE_f \varepsilon_1 - 2a\tau_{my} \coth(n\bar{s})]}{nE_f(v_1 E_m \varepsilon_1 - \sigma_r)}\right\} \qquad (5.59)$$

Since the interfacial shear stress depends strongly on the strain, ε_1, in this case, a critical aspect ratio cannot readily be determined. Thus the stress–strain relations have to be calculated using numerical methods.

If the strain at the slip point is such that $\sigma_{rt} > 0$ when slip occurs, reinforcement by slip cannot take place, so the specimen must fail at the slip point.

The fibres in poorly bonded reinforced metals will also suffer frictional slip instead of being stressed as a result of matrix shear. In this case the governing equations will be those used for reinforced ceramics.

Further Reading

Cox, H. L. (1952) *British J. Appl. Phys.* **3**, 72.
Outwater, J. (1956) *Modern Plastics* **33**, 37.
Kelly, A. and Davies, G. J. (1965) *Metall. Rev.* **10**, 1.
Piggott, M. R. (1978) *J. Mater. Sci.* **13**, 1709.

6

Failure Processes

So FAR in our discussion of stress transfer mechanisms we have stopped short at the ultimate tensile strain, or yield strain, of the components of the composite. We have also not considered the case where more than half the fibre length is subject to slip. These will be discussed in this section, but first we will examine the conditions for reinforcement to be obtained at all, and for multiple fracture to occur.

6.1. Critical Volume Fraction

If very few fibres are added to a matrix, the material is weakened rather than strengthened. For reinforcement we must have

$$\sigma_{1u} > \sigma_{mu}$$

where u in the subscripts denotes ultimate tensile strength.

As we have already shown, σ_{1u} should be a linear function of V_f, at least with well-made composites, with moderate values of $V_f (< 0.8)$. Thus there will be some positive value of V_f where

$$\sigma_{1u} = \sigma_{mu}. \tag{6.1}$$

This defines the critical volume fraction. We can recognize three cases: (1) $\varepsilon_{fu} < \varepsilon_{my}$; (2) $\varepsilon_{my} < \varepsilon_{fu} < \varepsilon_{mu}$; (3) $\varepsilon_{mu} < \varepsilon_{fu}$.

We will consider composites with fibres that are not necessarily aligned, but have high aspect ratios ($s \gg s_c$). We will write

$$\sigma_1 = (\chi_1 V_f E_f + V_m E_m)\varepsilon_1 \tag{6.2}$$

where χ_1 is a factor less than unity which makes allowance for fibres which are not aligned in the direction of the applied stress.

(1) $\varepsilon_{fu} < \varepsilon_{my}$. The composite failure starts when $\varepsilon_1 = \varepsilon_{fu}$. Thus

$$\sigma_{1u} = (\chi_1 V_f E_f + V_m E_m)\sigma_{fu}/E_f. \tag{6.3}$$

For the critical volume fraction $\sigma_{1u} = \sigma_{mu}$. Inserting this into equation (6.3), writing V_{fmin} for the critical volume fraction, remembering that $V_f + V_m = 1$, and rearranging, gives

$$V_{fmin} = \frac{E_f \sigma_{mu} - E_m \sigma_{fu}}{(\chi_1 E_f - E_m)\sigma_{fu}}. \tag{6.4}$$

For aligned carbon-epoxy (data from Tables 1.1 and 3.2, for the stiff carbon) this comes to $V_{f\min} = 0.020$. This is much less than the amounts normally used. For fibres distributed randomly in a plane, $\chi_1 \simeq 0.38$, and $V_{f\min}$ is increased to 0.052, still somewhat less than the amounts normally used. However, for the case of fibres distributed randomly in space, $\chi_1 \simeq 0.2$. A plot of σ_{1u} vs. V_f for this case is shown in Fig. 6.1.

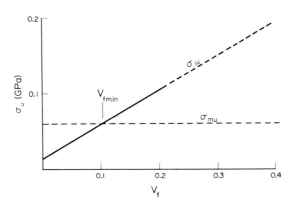

FIG. 6.1. Strength of random carbon-epoxy vs. V_f, showing critical volume fraction.

In this case $V_{f\min} = 0.10$: volume fractions less than this are quite common in these random fibre composites, because of the difficulty of packing large proportions of fibre. Thus strengthening is not always possible with random fibres.

(2) $\varepsilon_{my} < \varepsilon_{fu} < \varepsilon_{mu}$. The matrix yields when $\varepsilon_1 = \varepsilon_{my}$, but the fibres do not fail until $\varepsilon_1 = \varepsilon_{fu}$. At this strain the matrix will have yielded and work hardened if it is a metal. Polymers can show similar effects though this may not be apparent from the "engineering" stress–strain curves (Figs. 1.1 and 1.3). This is because the test piece narrows at one place, once yielding has taken place, and the stress and strain should be calculated on the basis of the reduced cross-section, rather than the original cross-section. Curves calculated from the reduced cross-section are called true stress-true strain curves. If we have such curves available we can determine the matrix stress at fibre failure, $\sigma_m(\varepsilon_{fu})$. Instead of equation (6.2) we now have

$$\sigma_{1u} = \chi_1 V_f \sigma_{fu} + V_m \sigma_m(\varepsilon_{fu}). \tag{6.5}$$

Putting $\sigma_{1u} = \sigma_{mu}$ and $V_f = V_{f\min}$ in this equation gives

$$V_{f\min} = \frac{\sigma_{mu} - \sigma_m(\varepsilon_{fu})}{\chi_1 \sigma_{fu} - \sigma_m(\varepsilon_{fu})}. \tag{6.6}$$

For the brass shown in Fig. 1.1 we can calculate $V_{f\min}$ because it did not neck. For reinforcement let us consider the strong carbon fibre, Table 3.2. $\varepsilon_{fu} = 0.012$, and thus $\sigma_m(\varepsilon_{fu}) = 150$ MPa while $\sigma_{mu} = 270$ MPa, so that $V_{f\min} = 0.045$ for $\chi_1 = 1$. For $\chi_1 = 0.38$, $V_{f\min} = 0.13$ and for $\chi_1 = 0.2$, $V_{f\min} = 0.21$. The critical volume fraction for the aligned case is again rather small, and seldom significant. For the random cases it is a very important parameter, however.

For the materials for which the true stress-true strain curves are not available σ_{my} may be used, instead of $\sigma_m(\varepsilon_{fu})$ to give a maximum bound for $V_{f\min}$. Note that for σ_{mu} the engineering stress should be used, and not the true stress at failure.

(3) $\varepsilon_{mu} < \varepsilon_{fu}$. We substitute $\varepsilon_1 = \varepsilon_{mu}$ into equation 6.2, which thus becomes

$$\sigma_1^* = (\chi_1 V_f E_f + V_m E_m)\sigma_{mu}/E_m \qquad (6.7)$$

where we have written σ_1^* for composite stress at this strain, since failure may not take place until the stress is much higher than this. Although the matrix should crack, the fibres can remain intact. The stress for the initiation of matrix cracking will equal the matrix strain at a fibre volume fraction V_{fc} where

$$\sigma_{mu} = (\chi_1 V_{fc} E_f + (1 - V_{fc})E_m)\sigma_{mu}/E_m$$

This is only true when $V_{fc} = 0$. The stress for matrix cracking exceeds σ_{mu} only when $\chi_1 E_f > E_m$; it then does so at all volume fractions.

Most aligned fibre composites do not fail at σ_1. The matrix cracks into small pieces, and the load is borne solely by the fibres. Then $\sigma_{1u} = V_f \sigma_{fu}$, and so V_{fmin} is given by $\sigma_{mu} = V_{fmin}\sigma_{fu}$, i.e.

$$V_{fmin} = \sigma_{mu}/\sigma_{fu} \qquad (6.8)$$

If a non-aligned fibre composite showed multiple fracture, we would expect V_{fmin} to be $1/\chi_1$ times greater than this. A typical value for aligned glass fibre reinforced plaster ($\sigma_{mu} = 7$ MPa, $\sigma_{fu} = 3.4$ GPa) is very small; 0.0021. For spatially random fibres V_{fmin} would be only about 1%.

Since low aspect ratio fibres give composites of reduced strength as compared with high aspect ratio ones, the critical volume fractions are correspondingly greater. (The reader can test this for himself using equations (4.15) and (4.16) for polymers, or equations (4.20) and (4.21) for metals.)

Note that a critical volume fraction does not exist for stiffening. Stiffening should occur at all volume fractions, with all fibre arrangements, providing $E_f > E_m$.

6.2. Multiple Fracture

When the component of a composite with the lower breaking-strain fails, complete fracture of the composite may not ensue. If the composite stress at that instant is less than the product of the strength of the higher failure strain component and its volume fraction, this component can bear the load without breaking. Multiple fracture of the other component then occurs. Figure 6.2 shows an example of multiple fibre fracture (c) and multiple matrix fracture (d). For simplicity we will consider high aspect ratio fibres ($s \gg s_c$). We recognize the same three cases as in the previous section, and will deal with each in turn.

(1) $\varepsilon_{fu} < \varepsilon_{my}$. In this case the fibres break first. We therefore compare the stress for the start of composite failure, given in equation (6.3), with $V_m \sigma_{mu}$. For multiple fibre fracture

$$V_m \sigma_{mu} > (\chi_1 V_f E_f + V_m E_m)\sigma_{fu}/E_f$$

or, rearranging

$$V_f < \frac{E_f \sigma_{mu} - E_m \sigma_{fu}}{(\chi_1 E_f - E_m)\sigma_{fu} + E_f \sigma_{mu}}.$$

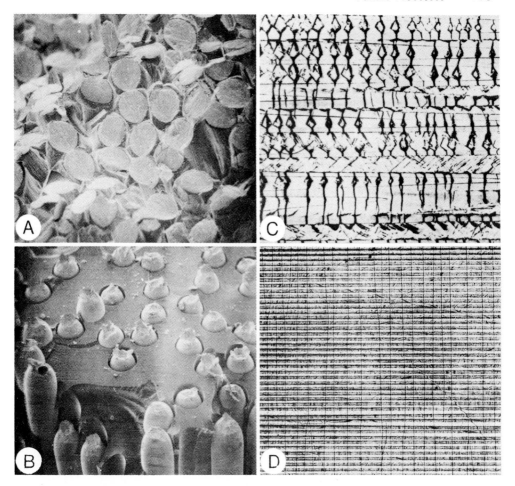

FIG. 6.2. Single and multiple cracking. (A) Single fracture, ductile matrix: W–Cu. (B) Single fracture, brittle matrix: phosphor bronze-epoxy. (C) Multiple fracture, ductile matrix: CoAl–Co. (D) Multiple fracture, brittle matrix: steel-epoxy 77K. (After Cline, H. E. (1966) GE Report 66-GC, and Cooper, G. A. and Piggott, M. R. (1977) *Fracture*, **1**, 557.)

This volume fraction is, of course, less than $V_{f\min}$. Thus we conclude that in this case multiple fibre fracture is an indication that too low a volume fraction was used.

(2) $\varepsilon_{my} < \varepsilon_{fu} < \varepsilon_{mu}$. This time we compare the composite stress in equation (6.5) with $V_m \sigma_{mu}$ since, again, the fibres break first. For multiple fibre fracture

$$V_m \sigma_{mu} > \chi_1 V_f \sigma_{fu} + V_m \sigma_m (\varepsilon_{fu})$$

or

$$V_f < \frac{\sigma_{mu} - \sigma_m(\varepsilon_{fu})}{\chi_1 \sigma_{fu} - \sigma_m(\varepsilon_{fu}) + \sigma_{mu}}.$$

Again, this volume fraction is less than $V_{f\min}$

(3) $\varepsilon_{mu} < \varepsilon_{fu}$. This time the matrix breaks first, so we compare $(\chi_1 V_f E_f + V_m E_m)\varepsilon_{mu}$

with $V_f \sigma_{fu}$. Thus multiple matrix failure occurs when

$$V_f \sigma_{fu} > (\chi_1 V_f E_f + V_m E_m) \sigma_{mu} / E_m$$

or rearranging

$$V_f > \frac{E_m \sigma_{mu}}{E_m \sigma_{fu} - \chi_1 E_f \sigma_{mu} + E_m \sigma_{mu}}. \tag{6.9}$$

This is only a little bit greater than σ_{mu}/σ_{fu}, so for aligned fibre composites multiple matrix failure occurs at volume fractions greater than or approximately equal to V_{fmin}. For glass cement $(\sigma_{mu} = 7$ MPa, $E_m = 20$ GPa, $\sigma_{fu} = 3.4$ GPa and $E_f = 72$ GPa) V_f must be greater than 0.0021 for multiple matrix failure. This important process will be examined in more detail in the next section.

6.3. Multiple Matrix Fracture

This is the only case of multiple fracture of importance where improved strength is required. (As shown in the previous section, when multiple fibre fracture is observed, the composite is weaker than the matrix.) For this $\varepsilon_{mu} < \varepsilon_{fu}$.

Multiple fracture is a progressive process, controlled by flaws in the matrix. The first crack occurs at the largest flaw. The pieces of matrix on either side of the crack, having smaller flaws, are stronger than the original piece of matrix, so a greater stress is required to cause further fracture. The fracture process proceeds until the matrix has broken into such small pieces that the fibres can no longer transmit the breaking-stress to them.

In order to understand the processes taking place we need to make a number of simplifying assumptions. We will neglect elastic stress transfer. This is only important in so far as it ensures that the matrix stress is always largest at the mid-points between adjacent fracture surfaces, thus making further cracking most likely to occur in the centre. Each piece therefore keeps dividing roughly into two equally sized pieces. We will assume that stress transfer takes place by friction, at constant shear stress τ_i, and that the matrix strength remains approximately constant.

We will first look at the situation before the first matrix crack appears. Equilibrium requires that

$$\sigma_1 = V_f \sigma_f + V_m \sigma_m \tag{6.10}$$

and $\varepsilon_f = \varepsilon_m = \varepsilon_1$, so that the stress–strain behaviour of the composite is given by $\sigma_1 = E_L \varepsilon_1$ where

$$E_L = V_f E_f + V_m E_m. \tag{6.11}$$

As the strain is increased σ_m reaches σ_{mu} before σ_f reaches σ_{fu}. Thus the first crack appears when

$$\sigma_1 = V_f \sigma_f + V_m \sigma_{mu}. \tag{6.12}$$

Further cracks are generated as the strain is further increased, and Fig. 6.3 is a schematic drawing of the fibre and matrix stresses while cracks are forming. Between adjacent crack faces there are three distinct regions. In the centre, over a distance $2L_m(1-m)$, the fibre and matrix stresses are constant, having the values σ_{fi} and σ_{mi} respectively. Near each crack

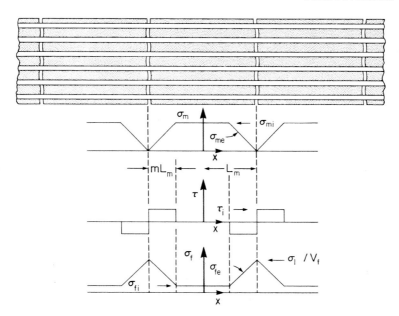

F_{IG}. 6.3. Multiple matrix fracture, and associated matrix, interface, and fibre stresses.

face, over a distance mL_m, slip is taking place, and

$$\frac{d\sigma_{fe}}{dx} = \frac{2\tau_i}{r}$$

(compare with equation (4.1)). Integrating this equation, with the boundary condition $\sigma_1 = V_f \sigma_f$ at $x = L_m$ (i.e. $\sigma_m = 0$ at crack faces) we find that

$$\sigma_{fe} = \frac{\sigma_1}{V_f} - \frac{2\tau_i}{r}(L_m - x) \tag{6.13}$$

At $x = L_m(1 - m)$, $\sigma_{fe} = \sigma_{fi}$ so that

$$\sigma_{fi} = \frac{\sigma_1}{V_f} - 2\tau_i m s_m \tag{6.14}$$

The matrix stress, σ_{mi}, is obtained by using equation (6.10) with $\sigma_m = \sigma_{mi}$ and $\sigma_f = \sigma_{fi}$, σ_{fi} being given by equation (6.14):

$$\sigma_{mi} = 2V_f \tau_i m s_m / V_m \tag{6.15}$$

After a matrix crack has appeared, there will be a sharp decrease in applied stress, and it will gradually increase again with increasing strain, as the slip region develops adjacent to the new crack faces. When $\sigma_{mi} = \sigma_{mu}$, a new crack appears. In this case $m = m_c$ where

$$m_c = \frac{V_m \sigma_{mu}}{2V_f \tau_i s_m} \tag{6.16}$$

and σ_1 has regained its value given by equation (6.12). This process involves s_m decreasing to a half of its previous value at each step. This can only continue until $m_c = 1$. Then equation (6.16) gives

$$s_m = \frac{V_m \sigma_{mu}}{2 V_f \tau_i}$$

so that the smallest value of s_m is a half of this, i.e.

$$s_{min} = \frac{V_m \sigma_{mu}}{4 V_f \tau_i} \qquad (6.17)$$

In practice, the matrix will break into blocks having lengths between $2rs_{min}$ and $4rs_{min}$. In the centre region ($x < L(1-m)$) the fibre and matrix strains are the same, since no slip is taking place. Thus, at the instant a new crack is formed, $\sigma_{fi} = \sigma_{mu} E_f / E_m$, and the composite stress is given by

$$\sigma_1 = E_L \sigma_{mu} / E_m. \qquad (6.18)$$

(This is equation (6.12) with $\sigma_f = \sigma_{mu} E_f / E_m$). The stress to cause cracking is thus constant.

The total fibre displacement is equal to the composite displacement, so we can calculate the composite strain from the fibre average strain, i.e. $\varepsilon_1 = \bar{\varepsilon}_f$. Consider one block of matrix between adjacent cracks. The fibre displacement, u_f, is given by

$$u_f = \frac{1}{E_f} \int_0^{L_m(1-m)} \sigma_{fi} dx + \frac{1}{E_f} \int_{L_m(1-m)}^{L_m} \sigma_{fe} dx. \qquad (6.19)$$

The average fibre strain is $\bar{\varepsilon}_f = u_f / L_m$. Thus substituting σ_{fe} from equation (6.13) and σ_{fi} from equation (6.14), integrating, and dividing by L_m, we have

$$\varepsilon_1 = \frac{1}{E_f} \left\{ \frac{\sigma_1}{V_f} - 2\tau_i m s_m \left(1 - \frac{m}{2} \right) \right\}.$$

When $m = m_c$ this equation gives a series of values of ε_1 which increase as s_m decreases. Substituting $m = m_c$ using equation (6.16), and using equation (6.18) we have

$$\varepsilon_1 = \varepsilon_{mu} \left(1 + \frac{V_m^2 \sigma_{mu} E_m}{4 V_f^2 \tau_i s_m E_f} \right). \qquad (6.20)$$

Cracking ceases when s_m becomes less than $2s_{min}$. Substituting for $2s_{min}$ using equation (6.17), we find that the maximum value for ε_1, marking the end of the region of constant stress, is

$$\varepsilon_{1max} > \varepsilon_{mu} \left(1 + \frac{V_m E_m}{2 V_f E_f} \right).$$

Since the crack spacing cannot be less than s_{min}, the upper bound for ε_{1max} is $\varepsilon_{mu}(1 + V_m E_m / V_f E_f)$. Thus

$$\varepsilon_{mu} \left(1 + \frac{V_m E_m}{2 V_f E_f} \right) \leq \varepsilon_{1max} \leq \varepsilon_{mu} \left(1 + \frac{V_m E_m}{V_f E_f} \right). \qquad (6.21)$$

The matrix cannot be affected by any further increase in the strain. The fibres can still extend however, so beyond this point the stress–strain curve has a slope of $V_f E_f$. The composite finally fails when $\sigma_1 = V_f \sigma_{fu}$.

The stress–strain curve thus has three distinct parts (Fig. 6.4). The low stress region, OA, has a slope E_L, and terminates when $\varepsilon_1 = \varepsilon_{mu}$. The next region, AB, has zero slope, and serrations resulting from the sudden appearance of matrix cracks. The third region, BC, has a slope of $V_f E_f$ and terminates when $\sigma_1 = V_f \sigma_{fu}$.

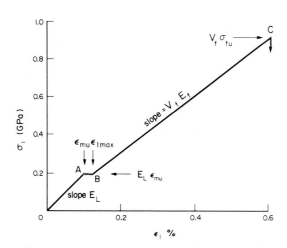

FIG. 6.4. Theoretical stress–strain curve for continuous aligned carbon–glass. $V_f = 0.4$.

In practice the region AB does have a positive slope because the matrix strength, σ_{mu}, increases as s_m decreases. In addition, an interesting synergism has been observed. When very fine fibres are used at high volume fractions ($V_f > 0.5$) the first crack appears at a strain greater than ε_{mu}. Thus fine fibres can inhibit matrix cracking.

The length of the region AB is proportional to $V_m E_m / V_f E_f$, so can be much longer than indicated in Fig. 6.4. For example, for glass-cement $V_m E_m / V_f E_f$ is 2.5 with $V_f = 0.1$ (as compared with 0.29 in Fig. 6.4). However, for this glass-cement the region BC is extremely long, making the region OAB appear insignificant.

When short fibres are used, the strength of the composite is decreased and for $s > s_c$ (where $s_c = \sigma_{fu}/2\tau_i$) the composite strength is

$$\sigma_{1u} = V_f \sigma_{fu}(1 - s_c/2s).$$

For $s < s_c$ the composite fails by fibre pull-out after matrix cracking and the breaking-stress is

$$\sigma_{1u} = V_f \tau_i s.$$

When the fibres are randomly oriented the strength of the composite is further reduced. For the planar random case σ_{1u} is reduced by a factor $3/8$, and for the spatially random case by $1/5$, as discussed in Section 4.3.4.

6.4. Yielding Matrices

With most metal–fibre combinations $\varepsilon_{my} < \varepsilon_{fu}$. When $\varepsilon_1 = \varepsilon_{my}$ the matrix yields, but the composite does not fail, since the fibres are not at their breaking strain. When $\varepsilon_1 > \varepsilon_{my}$ the matrix stress ceases to be $\bar{\sigma}_m = E_m \varepsilon_1$. Instead, for insertion into the Rule of Averages (equation (4.5)) we use $\bar{\sigma}_m = \sigma_{my}$. We are thus continuing to assume elastic-perfectly plastic behaviour. (It was assumed for the evaluation of matrix slip effects.)

When $\varepsilon_1 > \varepsilon_{my}$ stress transfer can still take place in the fibre centre region, $x < L(1-m)$, but with reduced shear stresses in the matrix. This is allowed for by using n^* instead of n (see equation (5.8)) where

$$n^* = n\sqrt{(E_m^{\prime*}/E_m)}$$

E_m^* is the scant modulus. At any strain $\varepsilon_m > \varepsilon_{my}$ (Fig. 6.5), $E_m^* = \sigma_{my}/\varepsilon_m$, where $E_m^* = \tan\phi$. Thus

$$n^* = n\sqrt{(\varepsilon_{my}/\varepsilon_1)}. \tag{6.22}$$

These changes mean that, for $\varepsilon_1 > \varepsilon_{my}$, instead of equation (5.36) we use

$$\sigma_1 = V_f E_f \varepsilon_1 + V_m \sigma_{my} - \frac{V_f}{2s}\left(\frac{E_m \varepsilon_1}{n^2} + \frac{E_f^2 \varepsilon_1^2}{\sigma_{my}}\right) \tag{6.23}$$

where we have written σ_{my} for $2\tau_{my}$ in the last term, in addition to the other changes.

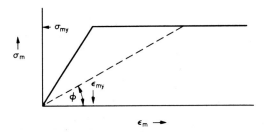

FIG. 6.5. Stress–strain curve assumed for a metal matrix. (Compare with Fig. 1.1.)

Adoption of this procedure reproduces the "knee" observed in stress–strain curves of reinforced metals. An example is given in Fig. 6.6 for boron–aluminium. Below the knee $E_1 = V_f E_f + V_m E_m$ for continuous fibres; above the knee the slope of the stress–strain curve is reduced to $V_f E_f$. The curves have been terminated at $\varepsilon_1 = \varepsilon_{fu}$. 0.25 was chosen for V_f in order to accentuate the knee. Note the small elastic region.

Fibre yielding before fracture has been observed with reinforced metals, most notably with tungsten reinforcement. In this case the stress–strain curve has three regions with different slopes. With continuous fibres these are: (1) $V_f E_f + V_m E_m$ near the origin; (2) $V_f E_f$ when $\varepsilon_1 > \varepsilon_{my}$; and (3) $V_f E_f^*$ when $\varepsilon_1 > \varepsilon_{fy}$. E_f^* is the secant modulus of the fibres. Some synergism has also been noted, so that composite failure takes place at strains greater than ε_{fu} with continuous fibres.

6.5. Single Fracture

This normally occurs with fibre reinforced metals and polymers, so long as $V_f > V_{fmin}$. With metals ε_{fu} is usually greater than ε_{my}, while with polymers it is less. We will consider the two cases separately in this section, for fibres with $s > 2s_c$.

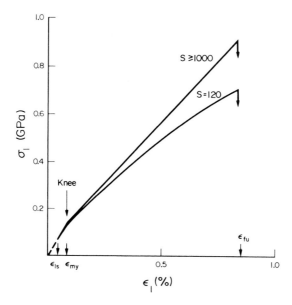

FIG. 6.6. Theoretical stress–strain curves for B–Al. $V_f = 0.25$.

6.5.1. $\varepsilon_{my} < \varepsilon_{fu} < \varepsilon_{mu}$

In this case the composite breaks when ε_1 reaches ε_{fu}, while the matrix is yielding. Thus we put ε_1, in equation (6.23), equal to ε_{fu}, and $\sigma_1 = \sigma_{1u}$. We replace $E_f\varepsilon_{fu}$ by σ_{fu} and σ_{fu}/σ_{my} by s_c, and so obtain

$$\sigma_{1u} = V_f\sigma_{fu}(1 - s_c/2s) + \sigma_{my}(V_m - V_f/2n^{*2}s) \qquad (6.24).$$

Remembering that $\sigma_{my} = \sigma_{mu}$ for an elastic-perfectly plastic material, we note that this reduces to the Rule of Mixtures expression for strength (equation (4.19)) when $s \to \infty$. This type of material does not at the same time obey the Rule of Mixtures for modulus (equation (4.12)) except at low stress, where $\varepsilon_1 < \varepsilon_{my}$.

Figure 6.6 shows stress–strain curves for boron–aluminium which terminate at ε_{fu}.

6.5.2. $\varepsilon_{fu} < \varepsilon_{my}$.

In this case the matrix is elastic up to the composite breaking-point given by $\varepsilon_1 = \varepsilon_{fu}$. Inserting ε_{fu} into equation (5.51) and replacing $E_f\varepsilon_{fu}$ by σ_{fu} and σ_1 by σ_{1u} we have

$$\sigma_{1u} = V_f\sigma_{fu} + V_mE_m\varepsilon_{fu} - \frac{V_f}{s}\left\{\frac{\sigma_{fu}^2 - (2a\tau_{my}/n)^2}{4\mu(v_1E_m\varepsilon_{fu} - \sigma_r)} + \frac{2a\tau_{my}}{n^2}\right\} \qquad (6.25).$$

Notice that this does not reduce to the Rule of Mixtures as $s \to \infty$. Usally the matrix contribution is small, so that the difference between this and the Rule of Mixtures is not important.

(Note that in this, and the previous case (6.5.1) we have assumed that coth $(n\bar{s}) = 1.00$. If this is not so, ε_{1u} slightly exceeds σ_{fu}; see equation 5.25)

6.6. Slip Failure

When the fibre aspect ratio is less than s_c the matrix cannot transfer enough stress to the fibres to break them. Instead the composite fails by the matrix slipping past the fibres, with $m \simeq 1$.

The stress–strain curves for short fibres ($s < s_c$) have three regions. First there is an elastic region governed by equation (5.13). This terminates at the slip point, given by equation (5.17) or (5.39) (for polymers), and slip starts. During slip we cannot assume $\coth{(n\bar{s})} = 1$. So the full equations ((5.32) and (5.33) or (5.47) and (5.49) together with the Rule of Averages, equation (4.5)) must be used. These require numerical methods for their solution. This second region ends when m reaches about 0.99. During the third region the average fibre stress, $\bar{\sigma}_f$, is independent of strain, ε_1.

(Composites having fibres with $s_c < s < 2s_s$ do not fail by slip; however, they must also be analysed using the full equations because we cannot assume $\coth{(n\bar{s})} = 1.00$).

We will first consider reinforced metals which slip at constant shear stress. Although $\coth{(n\bar{s})}$ goes to infinity as m approaches 1.0, the term involving it becomes insignificant. For $n\bar{s} \ll 1$, $\coth{(n\bar{s})} \simeq 1/n\bar{s}$. Substituting this in equation (5.33) and rearranging gives

$$2\tau_{my} s (1 - m)m \simeq E_f \varepsilon_1 (1 - m) - 2\tau_{my} n^2 s.$$

Substituting this into equation (5.32) gives

$$\bar{\sigma}_f = 2\tau_{my} s(1 - m)\, m + m^2 s \tau_{my}$$

which for $m \to 1$ reduces to

$$\bar{\sigma}_f = s\tau_{my} \tag{6.26}$$

so that $\bar{\sigma}_f$ is independent of ε_1, no matter how large ε_1 may be.

The strains at which slip failure starts is given approximately by ε_1 evaluated from equation 5.33, without the hyperbolic terms, and with $m \simeq 1$. Writing ε_{1p} for this we have

$$\varepsilon_{1p} \simeq 2\tau_{my} s / E_f. \tag{6.27}$$

We use the Rule of Averages (equation 4.5) for the composite stress, thus

$$\sigma_1 = V_f s \tau_{my} + V_m \bar{\sigma}_m. \tag{6.28}$$

For $\bar{\sigma}_m$ we use $E_m \varepsilon_1$ or σ_{my} according to whether or not the matrix yield stress is exceeded. Stress–strain curves for boron–aluminium developed using this treatment are shown in Fig. 6.7. They should be compared with the curves shown in Fig. 6.6 for $s > 2s_c$.

The corresponding equations for frictional slip (reinforced polymers) are

$$\bar{\sigma}_f = \mu s(v_1 E_m \varepsilon_1 - \sigma_r) \tag{6.29}$$

(this is equation (5.46) with $\bar{\sigma}_f = \bar{\sigma}_{fe}$ and $m = 1$) and

$$\varepsilon_{1p} = -2\mu s \sigma_r / (E_f - 2\mu s v_1 E_m) \tag{6.30}$$

(this is equation (5.49) rearranged, with $m = 1$ and $a = 0$).

Using the Rule of Averages (equation (4.5)) for the composite stress we thus get

$$\sigma_1 = V_f \mu s(v_1 E_m \varepsilon_1 - \sigma_r) + V_m \bar{\sigma}_m \tag{6.31}$$

where, again, $\bar{\sigma}_m = E_m \varepsilon_1$ or σ_{my} according to the value of ε_1.

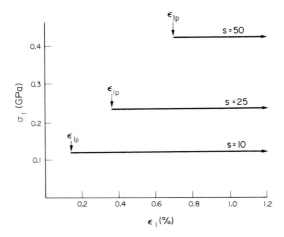

FIG. 6.7. Part of theoretical stress–strain curves for B–Al with $s < s_c$. $V_f = 0.25$.

Figure 6.8 shows stress–strain curves for carbon-epoxy for $s < s_c (= -\sigma_{fu}/2\mu\sigma_r)$. This should be compared with Fig. 5.8 which shows the stress–strain curves for the same material with larger aspect ratio fibres.

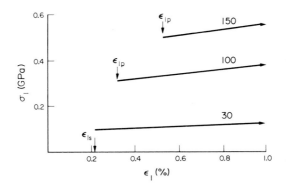

FIG. 6.8. Part of theoretical stress–strain curves for C-epoxy, with aspect ratios marked on curves, for $m = 1$. $V_f = 0.5$. (Note: $\varepsilon_{1s} \simeq \varepsilon_{1p}$ for $s = 30$.)

6.7. Factors Affecting Stress–Strain Relations

We have drawn theoretical stress–strain curves with approximate equations (Figs. 4.4, 4.6, and 4.7), which ignore elastic effects completely. We have also drawn them using a fairly complete analysis, but with equations derived for $\coth(n\bar{s}) = 1.00$, which is equivalent to assuming that the elastic contribution is independent of m (Figs. 5.6, 5.8, and 6.6).

Complete neglect of elastic effects gives average fibre stresses which are too high, and is roughly equivalent to using $s(1 + 1/n^2 s_c^2)$ instead of the true value of s. The error is thus usually not very great. However, an important feature in the theoretical curve is missed completely. This is the yield drop that is predicted for short-fibre-polymers.

With the more complete analysis, we are usually justified in assuming that the term involving coth $(n\bar{s})$ is either 1 or 0 according to whether $m < 0.99$ or $m > 0.99$. For example, Fig. 6.9 shows a set of stress–strain curves for B–Al making these approximations. They have a slight discontinuity, for $s < s_c$, when m reaches 1.0. This discontinuity is smoothed out if no approximation is used. Otherwise the curves are indistinguishable.

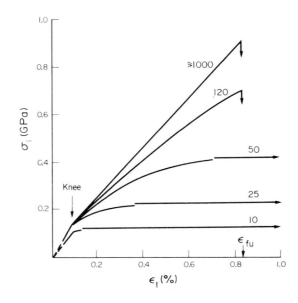

FIG. 6.9. Theoretical stress–strain curves for B–Al. $V_f = 0.25$, aspect ratios marked on the curves.

In the case of reinforced polymers, making these assumptions produces a much larger discontinuity for $s < s_c$. This, again, is smoothed out by eschewing approximations, as is illustrated in Fig. 6.10. The yield drop is still indicated, however, when $s < s_c$. For $s > s_c$, or at a distance from ε_{1p}, the curves are indistinguishable.

With reinforced ceramics such simple analysis is not possible since fibre Poisson's shrinkage becomes significant. Numerical methods then must be used.

Very few experiments have so far been carried out with well-characterized short ($s < s_c$) aligned fibre composites. This is unfortunate because the reinforcement and failure processes discussed here are fundamental to the whole technology of fibre reinforcement. One serious consequence of this is an almost complete lack of understanding of the role of fibre–matrix adhesion.

Good adhesion is not necessary for reinforcement. Slip occurs when $\varepsilon_{1s} \simeq 2\tau_{mu}/nE_f$ when adhesion is perfect ($\tau_{mu} =$ matrix shear strength). At this strain the fibre stress has only reached a small fraction of its fracture stress. Figure 6.11 shows the effect of varying the adhesion parameter, a, in carbon-epoxy at three aspect ratios. The feature most affected is the slip point, otherwise the stress–strain curves are little different, except for the disappearance of the yield drop for $s < s_c$.

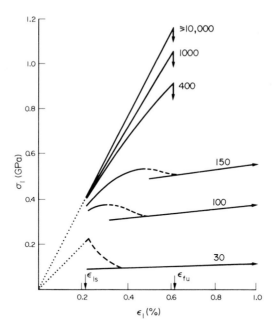

FIG. 6.10. Theoretical stress–strain curves for carbon-epoxy. Dashed curves indicate correct values when there is significant deviation due to the assumption that coth $(ns) = 1.00$. Aspect ratios marked on curves. Linear elastic region only included for lowest and highest aspect ratios. $V_f = 0.5$.

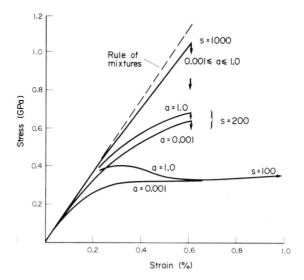

FIG. 6.11. Effect of adhesion coefficient, a, on stress–strain curves for carbon-epoxy. (For $s = 1000$ the curves for $a = 1.0$ and $a = 0.001$ are coincident). $V_f = 0.5$.

Reinforcement depends on slip, and it is essential that composites be designed to promote appropriate slip processes. Polymers and pure metals like aluminium are soft and yielding, so slip can take place relatively easily. However, this is not so with hard metals and ceramics. Much more attention needs to be directed towards slip promotion in these cases, especially when thermal cycling is likely to occur in addition to stress cycling. Successful whisker reinforced materials will also depend strongly on provision for appropriate slip mechanisms.

Good adhesion, however, is desirable for good properties in other than the fibre direction (see e.g. Section 6.8). It is also important for inhibiting environmental effects, such as occur when glass-polymers are immersed in water.

The coefficient of friction is very important in reinforced polymers. Hence carbon fibres are roughened before use. Interfacial layers providing desirable frictional properties offer a more rational way of controlling this, especially with short-fibre reinforced materials. Figure 6.12 shows the effect of varying the coefficient of friction for carbon-epoxy where the fibre aspect ratio is 200.

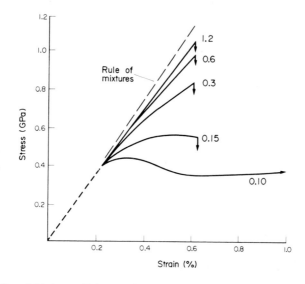

FIG. 6.12. Effect of friction coefficient on the stress–strain curve for carbon-epoxy for $s = 200$ and $V_f = 0.5$. The solid curves start at the slip point and the friction coefficients are marked on the curves.

Favourable shrinkage stresses are also very important for reinforced polymers and ceramics. Figure 6.13 shows that the yield drop can be very serious in the absence of compressive residual stresses. Tensile residual stresses can be disastrous. The critical aspect ratio becomes indefinitely large and failure at the slip point is indicated for short-fibre composites when tensile residual stresses exceed $2v_1 a\tau_{mu}E_m/nE_f$. (This is because σ_r must not exceed $v_1 E_m \varepsilon_{1s}$, else the fibres and matrix separate at the slip point.) The combination of no adhesion and no residual compression is particularly serious. Such a composite would be weaker than the matrix. These factors are discussed in more detail in Chapter 11, which describes reinforced ceramics.

A better understanding of all these processes awaits an evaluation of the Poisson's

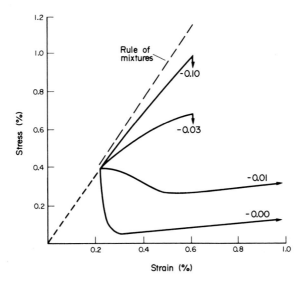

FIG. 6.13. Effect of residual stress on the stress–strain curve for carbon-epoxy for $s = 200$ and $V_f = 0.5$. The solid curves start at the slip point and the residual stresses (GPa) are marked on the curves.

shrinkage constants, v_1 and v_2. For the treatment here it has been adequate to assume that $v_1 = v_m$ (or, better, $v_1 = v_m/(1 + v_m)$, as has been done for the more accurate theoretical curves shown in Figs. 6.10–6.13). For reinforced ceramics we can use $v_2 = v_f$. However, when $v_1 \simeq v_2$ the errors in making these assumptions become substantial.

Table 6.1 gives values of Poisson's ratios for fibre and matrix materials.

TABLE 6.1. *Poisson's Ratios for Various Fibre and Matrix Materials*

Metal	v	Polymer	v	Ceramic	v
				Alumina	0.20
Aluminium	0.345	Epoxy	0.34	Boron	0.21
Copper	0.343	Kevlar[†]	0.35	Carbon	0.35[†]
Iron (steel)	0.287–0.295	Nylon	0.33	Cement	0.26
Magnesium	0.291	Polycarbonate	0.37	Glass	0.22
Molybdenum	0.293	Polyester	0.34	Silica	0.17
Nickel	0.293	Polyimide	0.33	Silicon	0.27
Tungsten	0.280	Polystyrene	0.33	Silicon carbide	0.19

[†] Transverse shrinkage of fibres.

6.8. Transverse Failure

Aligned fibre composites are very weak when stressed normal to the fibre axis. This is because the fibres usually do not contribute to the strength, and the fibre–matrix interface is often weak. If σ_i is the interfacial tensile strength, imperfectly bonded composites have a transverse strength governed by a Rule of Mixtures expression:

$$\sigma_{2u} = V_f\sigma_i + V_m\sigma_m. \tag{6.32}$$

(This result follows from similar arguments as those used to derive the Rule of Averages, equation (4.5)). It should be noted that σ_i is not equal to twice the τ_i used previously. When the interface is very strong, such that $\sigma_i > \sigma_{mu}$, then

$$\sigma_{2u} = \sigma_{mu}. \tag{6.33}$$

(This is an upper bound for the strength, difficult to achieve because of the stress concentrations created by the fibres.) In either case $\sigma_{3u} = \sigma_{2u}$.

In the case of fibres which can easily be split transverse to the applied stress (this happens sometimes with boron fibres) where the transverse fibre strength, σ_{f2u}, is less than both σ_i and σ_{mu}, then

$$\sigma_{2u} = V_f \sigma_{f2u} + V_m \sigma_{mu} \tag{6.34}$$

$\sigma_{3u} = \sigma_{2u}$ unless this fibre weakness has some preferred orientation.

Since the transverse strength is so low and the modulus, E_2, is relatively high (equation (4.22)) the breaking-strain is usually low. For E-glass-polyester, for example, it is often as little as 0.4%. This is less than the longitudinal failure strain, ε_{1u}. This is why in cross-ply laminates the first failure is observed in the laminae with the fibres normal to the stress. This leads to leakage before complete failure in fibreglass pressure vessels. Some alleviation of this problem results from the use of high breaking-strain, low modulus, matrices.

6.9. Shear Failure

The shear strength is also governed by the matrix and interface strengths, and is thus small. Because of this, deep beams made from aligned fibre composites tend to fail in shear when flexed.

Three principal shear strengths may be recognized (Fig. 6.14). If the shear strength of the fibre–matrix bond is τ_i, the forces are in equilibrium at the breaking-point when

$$\tau_{12u} = BV_f \tau_i + V_m \tau_{mu} \tag{6.35}$$

For an aligned fibre composite the factor B has been introduced to allow for the extra surface area of the fibre–matrix interface. Since the fibre profile is circular, $B \simeq \pi/2$ for regular packing of the fibres.

If the fibre–matrix bond is very strong, instead of equation (6.35) we have

$$\tau_{12u} = \tau_{mu}. \tag{6.36}$$

This is an upper bound for the composite shear strength.

The other in-plane shear strength τ_{13u}, should obey the same equations. However, the out-of-plane shear strength, τ_{23u}, is greater than that given by equation (6.35) because this shear does not involve sliding parallel to the fibre axes. τ_{23u} cannot exceed τ_{mu}, however.

6.10. Compressive Failure

The compressive strength is important because composites are often used as beams, and thin composite beams frequently fail first on the compressive side.

Rosen was the first to attempt to explain the compressive strength (see the reading list at

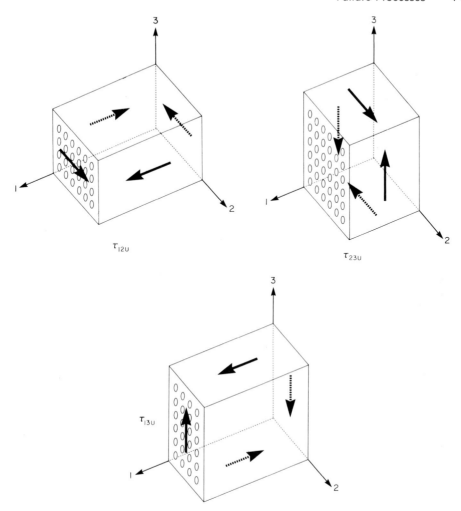

FIG. 6.14. The three principal shear-failure strengths of aligned fibre composites.

the end of the chapter). His theory was based on elastic buckling of the fibres, and produced a result of the form

$$\sigma_{1cu} = E_m/2V_m(1 + v_m). \tag{6.37}$$

This is far too high, especially for reinforced metals. It may be regarded as an upper bound for the strength in much the same way that the theoretical strength (equation (2.8)) may be regarded as an upper bound for the tensile strength of traditional materials. Note that both are fractions of the Young's modulus.

Before this strength can be achieved, other failure modes supervene. For example, Hayashi and Koyama suggest that matrix yielding initiates failure. Thus, for continuous fibre composites

$$\sigma_{1cu} = V_f\sigma_f + V_m\sigma_{my} \tag{6.38}$$

where the fibre stress, σ_f, is the stress achieved at the matrix yield strain. Consequently,

$$\sigma_f = \sigma_{my} E_f / E_m. \tag{6.39}$$

This gives the correct variation with V_f as observed experimentally: Fig. 6.15 shows the results obtained with tungsten–copper. Clearly these results do not fit the Rosen equation. Unfortunately, although equations (6.38) and (6.39) fit some results obtained with reinforced polymers quite well, they underestimate the strength of reinforced metals.

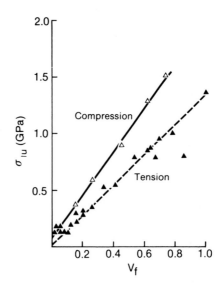

FIG. 6.15. Tensile and compressive strength of W–Cu. (After Kelley, A. (1973) *Strong Solids,* Clarendon Press, Oxford pp. (170–1.)

In the case of the tungsten–copper in Fig. 6.15 the elastic strain at the failure stress would be close to 0.5 %. This is much greater than the yield strain of the copper matrix and indicates that in this type of composite the compressive strength is given by a simple Rule of Mixtures expression:

$$\sigma_{1u} = V_f \sigma_{fcu} + V_m \sigma_{mu} \tag{6.40}$$

rather than equations 6.38 and 6.39. Here σ_{fcu} is the compressive strength of the fibres.

Similar results are obtained with steel wire reinforced polymers. Even when the steel is very hard and brittle equation (6.40) governs the compressive strengths of these composite when the wires are aligned, except that we use $E_m \varepsilon_{fu}$ instead of σ_{mu}, since $\varepsilon_{fu} < \varepsilon_{mu}$ (see e.g. equation (6.25): in practice this makes very little difference).

Kevlar fibres are much weaker in compression than in tension; the tensile strength is about eight times the compressive strength. In addition the modulus in compression is low; about' half the tensile modulus: it is ill-defined, and permanently reduced by compression at moderate stresses. Thus composites made with aligned Kevlar fibres have a strength given by equation (6.40) where $\sigma_{fcu} \simeq \frac{1}{8}\sigma_{fu}$ (tension).

The measurement of the compressive strength and the determination of failure mechanisms in aligned fibre composites are hampered by side effects occurring during the

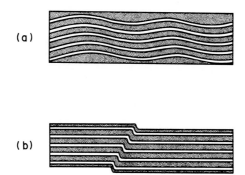

(a)

(b)

FIG. 6.16. (a) Elastic compressive buckling, (b) compressive failure. (After Hayashi, T. and Koyama, K. (1971) *Proc. Int. Conf. Mech. Behav. Mater.*, Kyoto, **5**, 104.)

compression test. These include: (1) the specimens tend to buckle, and because of the high degree of anisotropy, this occurs at stresses less than one third that expected for isotropic materials: (2) friction occurs at the specimen-machine interface, impeding Poisson's expansion there, so that the specimen tends to become barrel shaped; and (3) the specimens tend to split and open up like a brush at the ends.

FIG. 6.17. Compression failure of glass-epoxy pultruded rod.

Observation of reinforced polymers in compression shows that uniform buckling of the type shown in Fig. 6.16a does occur while the specimen is still elastic. However, when the specimen is properly supported at the ends failure occurs by sudden catastrophic buckling in one cross section, as shown in fig. 6.16b. Fig. 6.17 shows a pultruded rod that has failed in this way. (If testing is carried out in a very 'hard' machine the buckling can be made to progress across the specimen at a controlled rate). This type of failure is also observed with reinforced metals.

In practice, aligned fibre composites have very variable compressive strengths and moduli. When the modulus of the fibres is high (e.g. as it is for carbon) the compressive modulus can be substantially less than the Rule of Mixtures value. This is almost certainly due to lack of straightness of the fibres, and consequently when extreme care is taken to keep the fibres as straight as possible the properties are much improved.

The fibre-matrix adhesion is also important. Its effect is particularly marked when $V_f > 0.5$. Under compression the Poisson's ratio effect is reversed, and so there is usually a tensile stress at the interface. This may cause adhesive failure between fibres and matrix.

The role of the matrix is simply to support the fibres against buckling. This is clearly demonstrated by a series of experiments in which pultruded rods were tested at various stages of matrix cure. During the early stages of cure when the matrix was quite soft, the fibres buckled easily, and the strength was determined by the matrix yield stress, i.e.

$$\sigma_{1u} = B\sigma_{my}^q \tag{6.41}$$

where $q = 1$ and $B = 10$ for $V_f = 0.31$ and $q = 0.6$ and $B = 50$ for $V_f = 0.55$. The composite modulus also depended on the matrix yield stress, but once this exceeded 60 MPa, further increase in yield stress had no effect.

For aligned fibre composites compressed in the other two directions the strength is given by similar expressions to those used for the tensile strength (e.g. equation 6.33)

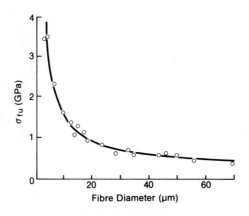

FIG. 6.18. Effect of fibre diameter on strength. (After Metcalf, A. G. and Schmitz, K. G. (1974) ASTM Proc. **64**, 1075.)

6.11. Statistical Aspects of Failure

Fibres and whiskers possess their uniquely high strengths by virtue of the suppression of imperfections; nevertheless, some imperfections still remain. The larger the diameter of the fibre the larger the imperfections can be, and hence the weaker the fibre becomes. Figure 6.18 shows typical results obtained with glass. With a given fibre diameter, increasing the length of a fibre increases the probability of a large flaw. Thus longer fibres are usually weaker than short ones. This was appreciated in 1939, and Fig. 6.19 shows some results obtained at that time.

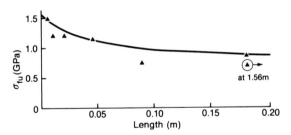

FIG. 6.19. Effect of fibre length on strength. (After Anderegg, F. (1939) *Ind. Eng. Chem.* **31,** 290.)

Imperfections are distributed at random and are of variable severity (the most severe ones usually being surface cracks) and hence the analysis of their influence on composite properties has to be carried out with the aid of statistics. Different 25-mm lengths cut from the same fibre have different strengths. This is illustrated in Fig. 6.20 which shows results from bending and tensile tests. Note that the bending tests, carried out on a very short length indeed (0.75 mm), give a higher average result, and are less variable than the tensile tests carried out on 25-mm lengths.

From the results shown in this figure, the average strengths and the coefficients of variation can be calculated.

With these data, we are able to calculate the strength of a bundle of fibres. The bundle strength will be much less variable than individual fibre strengths, but its average value will also be less. For example, for fibres having a coefficient of variation of 50%, the bundle strength is only half the mean fibre strength. However, in a composite, the fibres can

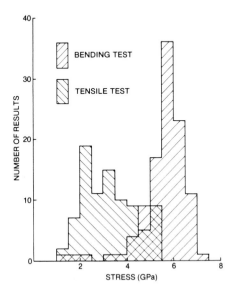

FIG. 6.20. Tensile and flexural strengths of fibres at 20 C. (After Piggott, M. R., and Yokom, J. (1968) *Glass Technol* **9,** 172.)

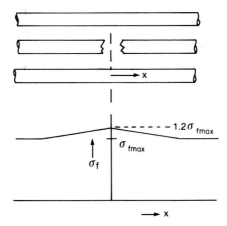

FIG. 6.21. Excess stress at a fibre break, hexagonal packing.

appear to be stronger than tests on individual fibres would suggest. This is because stress can be transferred from one fibre to another and back again, so that if one fibre fails, adjacent fibres will take a higher stress over a short distance close to the fracture point (Fig. 6.21). This distance will be about 0.2 rs_c. For a glass fibre, in a resin matrix exerting an interfacial shear stress of 10 MPa, this comes to only about 0.2 mm. It is most unlikely that an adjacent fibre has a serious enough flaw somewhere in this short length to cause it also to break, when taking on its share of the extra load due to the original fibre break. (Remember that the strength of the 0.74 mm length was about three times greater than the 25 mm length (Fig. 6.20).)

Rosen developed a statistical analysis of the situation and his results can be plotted as fibre efficiency factor, χ_2, defined by the expression:

$$\sigma_{cu} = \chi_2 V_f \sigma_{fu} + V_m \bar{\sigma}_m \tag{6.42}$$

The plot is reproduced in Fig. 6.22. When the length used for strength measurement is

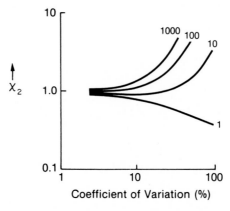

FIG. 6.22. Fibre efficiency factor vs. coefficient of variation for four different fibre test lengths. (After Rosen, B. W. (1964) *AIAA Journal*, **2**, 1985.)

short, i.e. the curve marked 1 on the figure, χ_2 is close to 1.0. For longer test lengths (the curves marked 10, 100, and 1000) χ_2 can be greater than one, especially when the coefficient of variation is large. However, a difficulty with the use of such plots is the paucity of data on flaw distribution.

There is also a graphical approach to the problem. Fibre strength is plotted against fibre length as in Fig. 6.23. Also plotted is the fibre stress when the fibre is being pulled out of the matrix. This varies with embedded length l_p, according to the expression

$$\sigma_{fp} = 2\tau_s l_p / r. \tag{6.43}$$

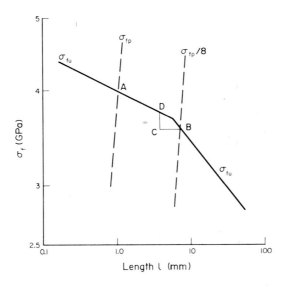

FIG. 6.23. Graphical method for determining the appropriate fibre strength in a composite. (Data for glass from Metcalf, A. G. and Schmitz, K. G. (1974) *ASTM Proc.* **64**, 1075.)

The intersection of the line for σ_{fp} with the fibre strength gives an upper limit for the strength to be expected from the reinforcing fibres. This is the point A on Fig. 6.23, and is plotted for $\tau_i = 10$ MPa and $2r = 10\ \mu$m. Fibres having a length represented by point A have the critical aspect ratio.

A more precise evaluation of the effective fibre strength can be obtained from the following procedure. Draw a line for $\sigma_f = \sigma_{fp}/8$. Let the intersection point, B (Fig. 6.23), have co-ordinates (l_B, σ_{fB}). The length l_B corresponds to $s \simeq 4s_c$.

Next draw the triangle BCD, where the length BD represents a reduction in 1 by a factor of two. (This is about 0.3 of the distance, along the l axis representing each decade.) The point D has co-ordinates $(l_B/2, \sigma_{fD})$. If

$$\sigma_{fB}/16 < \sigma_{fD} - \sigma_{fB} < \sigma_{fB}/8$$

the appropriate fibre stress is very close to σ_{fB}. If

$$\sigma_{fD} - \sigma_{fB} > \sigma_{fB}/4$$

we need a longer fibre length. The same procedure should then be adopted for $l = 2l_B$, but this time the limits for the inequality are $\sigma_{fB}/32$ and $\sigma_{fB}/16$. If

$$\sigma_{fD} - \sigma_{fB} > \sigma_{fB}/4$$

adopt the same procedure for $l = l_B/2$ using the limits $\sigma_{fB}/8$ and $\sigma_{fB}/4$.

This process is very unlikely to go beyond two iterations. For the glass shown in Fig. 6.23, and for Courtaulds carbon fibres, the slope of the logarithmic plot for fibre strength is such that the first step in the procedure gives the appropriate fibre strength with sufficient accuracy. (The result is slightly too high, while the next iteration gives a value which is too low, but the difference between them is only 4%. There is no point in trying to be more accurate than this, since the fibres are usually damaged when a composite is made, so that the composite strength is often much less than the theoretical value.)

The procedure described above is designed to determine the maximum value of $\bar{\sigma}_f$ using the equation

$$\bar{\sigma}_f = \sigma_{fu}(1 - s_c/2s)$$

(see equation (4.21), and remember that σ_{1u} is given by the Rule of Averages). In this equation both σ_{fu} and s_c are functions of the fibre length. It is assumed that when the composite is close to the breaking-point, the fibres break at their weakest points, and get progressively shorter, before the composite fails completely. Thus in the above case their aspect ratios are reduced to about $8s_c$. There is good evidence for early fibre fracture. Figure 6.24 shows a stress–strain curve for carbon-epoxy with the amount of acoustic emission at each strain superimposed. The emissions are considered to result from fibre failures, and they become very frequent close to the breaking-point.

The fibre shortening does not cause noticeable curvature in the stress–strain trajectory because it terminates when $s \simeq 8s_c$, at which point the composite fails catastrophically.

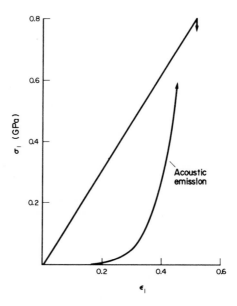

FIG. 6.24. Stress–strain curve and accompanying acoustic emission for carbon-epoxy. (After Fuwa, M., Ph. D. (1974) Thesis, Univ. of Bath.)

It is now clear why χ_2 in equation (6.41) can be greater than 1.0. Fibre strengths are normally measured on lengths of 25–50 mm. For a 50-mm length the average strength for the glass fibres shown in Fig. 6.23 is 2.85 GPa. The reinforcement strength corresponds to a length of about 7 mm, and is 3.6 GPa. Thus χ_2 will come to about 1.26 in this case.

Careful studies of continuous carbon-epoxies have shown that χ_2 can be considerably greater than 1.0 in this case also. These effects are not normally noticed, however, because of fibre damage during composite manufacture.

Further Reading

KELLY, A. and DAVIES, G. J. (1965) *Metall.* Rev. **10**, 1.
PIGGOTT, M. R. (1978) J. *Mater. Sci*, **13**, 1709.
COOPER, G. A. and KELLY, A. (1969) ASTM STP452 (ASTM, Philadelphia.)
ROSEN, B. W. *AIAA Journal* **2**, 1964, 1985.
PIGGOTT, M. R. and HARRIS, B. (1980) *J. Mater. Sci.*

7

Failure at Notches

FAILURE of notched specimens is governed by the toughness, or work of fracture, **G**, of a composite. We will discuss only one mode of fracture in this chapter, the opening mode. In this mode the stress is normal to the crack plane. There will, in general, be three principal works of fracture corresponding to stresses along the three axes in the composite. (The axes are shown in Fig. 4.8a.)

The inhomogeneity of the composite ensures that the work of fracture for stresses in the fibre direction of an aligned fibre composite, **G₁**, is not given by a Rule of Mixtures expression, i.e.

$$\mathbf{G}_1 \neq V_f \mathbf{G}_f + V_m \mathbf{G}_m.$$

Instead, the fibres and matrix interact in a number of different ways, sometimes co-operating to increase the work of fracture, and sometimes acting cohibitively.

This chapter discusses the main processes at work when a crack moves through a composite.

7.1. Short Fibres

We consider aligned fibres all having the same aspect ratio. Figure 7.1 shows the situation around the crack tip. The opening of the crack under the applied stress has caused fibres to pull out from both faces. To calculate the contribution of this process to **G**, consider one fibre (Fig. 7.2).

We assume that the interfacial shear stress, τ_i, is constant during fibre pull-out. When the fibre has an embedded length l, the force required to pull it out is $2\pi r l \tau_i$. The force exerted by the fibre is $\pi r^2 \sigma_f$, so that

$$\sigma_f = 2l\tau_i/r. \tag{7.1}$$

If $l/r > s_c (= \sigma_{fu}/2\tau_i)$, $\sigma_f > \sigma_{fu}$ and the fibre will break rather than pulling out. For the moment, consider a fibre with aspect ratio less than s_c. It will pull out, and while doing so σ_f will decrease linearly with the embedded length. The work of pull-out is

$$U_{fp} = \int_0^l 2\pi r x \tau_i dx$$

therefore

$$U_{fp} = \pi r l^2 \tau_i. \tag{7.2}$$

126

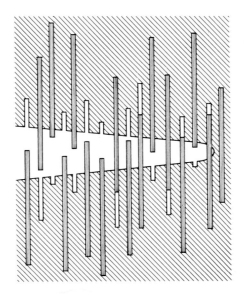

FIG. 7.1. Short fibres by a crack.

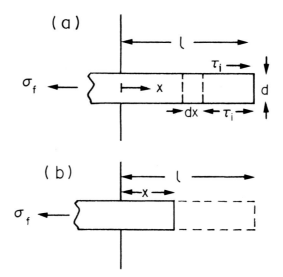

FIG. 7.2. Fibre pull-out stresses.

If there are N fibres crossing unit area of crack, the number with an embedded length between l and $l + dl$ on one side of the crack is $Ndl/2L$ where $2L$ is the fibre length. (We are assuming the fibres are randomly disposed, but parallel to each other.) The work done by these fibres in pulling out is $\mathbf{G}_{fp}/2$, since we are considering one side only of the crack. The work can be determined by summing the relevant values of U_{fp} for all the fibres. Thus

$$G_{fp}/2 = \int_o^L \frac{NU_{fp}dl}{2L}$$

so that the work of fracture for pull-out is

$$G_{fp} = V_f d\tau_i s^2/6 \qquad (7.3)$$

since $V_f = N\pi r^2$ and $s = L/r$. (d = fibre diameter = $2r$.)

This has its maximum value when $s = s_c$, which is

$$G_{fp\,max} = V_f d\sigma_{fu}^2/24\tau_i. \qquad (7.4)$$

For a carbon-epoxy with $\tau_i = 6$ MPa, $\sigma_{fu} = 2.3$ GPa, $d = 8$ μm, and $V_f = 0.50$ this comes to about 150 kJ m^{-2}, showing that, under ideal conditions, fibre pull-out can contribute a great deal to the work of fracture.

7.2. Continuous Fibres

With continuous fibres of uniform strength, fibre failure should occur in the crack plane, since that is where the fibre stress is greatest. The fibre is stretched during crack opening, sliding against the matrix and thus doing work. After it breaks it retracts into the matrix, losing its stored elastic energy.

The elastic energy stored in a short length, dx, of fibre is $\pi r^2 \sigma_f dx/2E_f$. Consider the elastic energy of the part of the fibre, length l_c, near the crack face, where slip has taken place (Fig. 7.3). Since the fibre eventually breaks, $l_c = rs_c$. We include only elastic energy arising from crack-induced stresses.

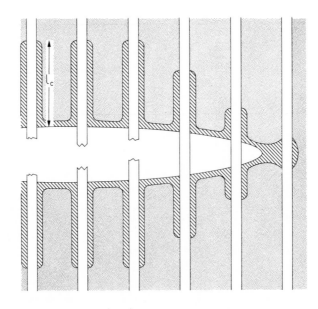

FIG. 7.3. Continuous fibres which fail in the crack plane as the crack opens. The fibre stretches and slides against the matrix over a distance l_c. Hatched regions indicate yielded matrix.

If x is the distance of the fibre element from the crack plane, then σ_f is given by equation (7.1) with $l_c - x$ replacing l. The elastic energy, dU_{fb}, of the element is

$$dU_{fb} = 2\pi(l_c - x)^2 \tau_i^2 dx/E_f. \tag{7.5}$$

The work done by the element in sliding against the matrix is

$$dU_{mf} = 2\pi r \tau_i u dx \tag{7.6}$$

where u is the displacement of the fibre element relative to the matrix, given by

$$u = \int_x^{l_c} \varepsilon_f dx.$$

(At $x = l_c$ the fibre strain, and displacement arising from the crack-induced stress, are zero.) ε_f, the fibre strain, can be determined from the fibre stress given in equation (7.1), with $l_c - x$ replacing l. Substituting for ε_f and integrating the equation thus obtained gives

$$u = \tau_i(l_c - x)^2/rE_f. \tag{7.7}$$

Substituting the expression for u into equation (7.6) shows that $dU_{mf} = dU_{fb}$. The total work is the sum of dU_{fb} and dU_{mf}, integrated between l_c and 0. Thus

$$U_{fb} + U_{mf} = \frac{2}{E_f} \int_{l_c}^{0} 2\pi \tau_i^2 (l_c - x)^2 dx.$$

The corresponding work of fracture, $\mathbf{G}_{fb}$, is equal to $2N(U_{fb} + U_{mf})$, where N is the number of fibres per unit area. Doing the above integration, and replacing l_c by $\sigma_{fu}d/4\tau_i$ and $N\pi r^2$ by V_f, we obtain

$$\mathbf{G}_{fb} = \frac{V_f d\sigma_{fu}^3}{6E_f \tau_i} \tag{7.8}$$

For the carbon-epoxy for which pull-out gave a work of fracture of about 150 kJ m^{-2}, $\mathbf{G}_{fb}$ comes to about 3.6 kJ m^{-2}. However, it is seldom practical to make aligned fibre composites with fibres having the critical aspect ratio (the carbon fibres in this example would only be 1.8 mm long), and both strength and modulus are degraded markedly. (Figs. 5.8 and 6.10 show stress–strain curves for this material when the fibres have various aspect ratios.)

7.3. Long Fibres

Consider fibres with $s \geqslant s_c$. As for pull-out, the number of fibres with embedded lengths between l and $l + dl$, on one side of the crack, is $Ndl/2L$. However, this time some fibres will break and others will not, according to the embedded length. To calculate the work of fracture in this case we must sum the appropriate amounts of U_{fp} and U_{fb}:

$$\tfrac{1}{2}\mathbf{G}_i = \frac{N}{2L} \int_0^{l_c} U_{fp} dl + \frac{1}{2L} \int_{l_c}^{2L - l_c} 2U_{fb} dl.$$

Making the appropriate substitutions (equations (7.2) and (7.8), remembering that

$\mathbf{G}_{fb} = 4U_{fb}$ and $V_f = N\pi r^2$) we obtain

$$\mathbf{G}_i = \frac{V_f d\sigma_{fu}{}^3}{6\tau_i}\left\{\frac{1}{8\tau_i s} + \frac{1}{E_f}\left(1 - \frac{\sigma_{fu}}{2\tau_i s}\right)\right\}. \tag{7.9}$$

A plot of $\mathbf{G}_i$ as a function of aspect ratio is shown in Fig. 7.4. The plot, presented in dimensionless form, shows that $\mathbf{G}_i$ has its maximum value at $s = s_c$. For $s \leqslant s_c$, $\mathbf{G}_{fp}$ has been plotted. The importance of fibre-breaking strain, for fibres with high aspect ratio, is clearly demonstrated.

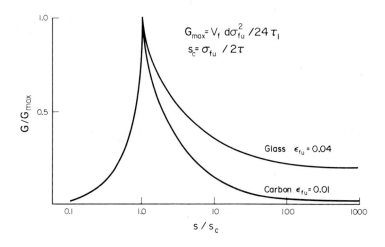

FIG. 7.4. Effect of aspect ratio on toughness.

7.4. Flawed Fibres

So far we have considered the ideal case of flaw-free fibres. Real fibres have many flaws in them, and this can be an advantage, so far as toughness is concerned.

Flawed fibres have a length–strength relationship such as that shown in Fig. 6.23. The strength of glass fibres with $s = s_c$, at an interfacial shear stress of 6 MPa, is about twice that of glass fibres of the normal length used to measure the strength (25 mm or more). If we now take Poisson's contraction into account we find that short lengths of glass fibre, at twice the apparently normal breaking-stress, can separate completely from the matrix.

This explains why some fibres can separate from the matrix near crack faces. Separation has been observed under the microscope, in the case of glass fibres bridging a slowly extending crack. The process started at the crack face, and extended along each fibre until it reached a suitably severe fibre flaw.

Since the fibre separated from the matrix, its stress must have been approximately constant along its length, between the crack face and the flaw. The fibre then broke at the flaw and pulled out.

The breaking of the fibre reduced the average stress in the fragment being pulled out, so its Poisson's contraction was less, and τ_i was no longer negligible. Thus the fibre fragment could dissipate a substantial amount of work while being pulled out.

This pull-out work, occurring with continuous fibres, can be much greater than the stress relaxation work described in Section 7.2. Low modulus fibres, like glass, contract much more than the higher modulus carbon and boron fibres. Thus separation from the matrix occurs much more readily with them. This is probably why glass-polymers are much tougher than carbon- and boron-polymers. (Kevlar-polymers are tough because Kevlar fibres have a high work of fracture. The works of fracture for glass, boron, and carbon fibres are all very small indeed.) Continuous carbon fibres can show pull-out effects, as shown in Fig. 7.5.

10 μ

FIG. 7.5. Fracture and pull-out of continuous carbon fibres. (Courtesy of Woodhams, R. T., University of Toronto.)

Without a detailed knowledge of the flaw distribution we cannot predict this pull-out work. We also need to know the residual radial stress at the fibre–matrix interface, and the coefficient of friction there. Data on all these parameters is very scanty indeed.

If, instead of using single fibres as reinforcement, we use bundles of fibres, we find that the bundles have uniform strength. Thus no substantial pull-out effects are obtained. However, even though the work of fracture comes mainly from stress relaxation, it can still be quite large. This is because bundles containing, say, 1000 fibres, will have an effective diameter of about thirty times the individual fibre diameter. The increased diameter increases $\mathbf{G}_{fb}$ so that it becomes comparable with $\mathbf{G}_{fpmax}$. Figure 7.6 shows that fibre bundles fail close to the crack plane.

Fig. 7.6. Macro and micro-graphs of glass- (*a* and *b*) and carbon- (*c* and *d*) polyester fracture surfaces.

7.5. Crack Opening

G_{fpmax}, G_{fb} and G_i can all be increased by increasing the fibre diameter and decreasing the interfacial shear stress. However, this is done at the expense of increasing the crack opening displacement.

The crack must open to a certain extent for the work of fracture to be dissipated. If the opening required is too large, complete failure of the matrix can occur, and the structure can fail, without the fracture work being dissipated. A good example of this is the reinforced concrete lamp post knocked over by a car. Complete matrix failure has occurred, the upper part of the post has fallen over, yet the steel reinforcement is still intact.

A figure of merit has been developed for this which we shall call the fracture integrity. This is the ratio of G to the crack opening displacement (COD).

In the case of pull-out work, for $s \leqslant s_c$, COD $= rs$. Using equation (7.3)

$$\frac{G_{fp}}{rs} = V_f \tau_i s/3. \tag{7.10}$$

For G_{fp}/rs to be as large as possible the aspect ratio should be large, and so should τ_i and V_f. At the maximum value of s, i.e. $s = s_c$ we have

$$\frac{G_{fpmax}}{rs_c} = V_f \sigma_{fu}/6. \tag{7.11}$$

In this case the fibre strength and volume fraction are the important factors.

For continuous fibres of uniform strength the criterion is $G_{fb}/2u$. For G_{fb} we use equation (7.8) and for u we use equation (7.7) with $x = 0$. Thus we get

$$\frac{G_{fb}}{2u} = 2V_f \sigma_{fu}/3. \tag{7.12}$$

Again, the important factors are fibre strength and volume fraction.

Comparing equations (7.11) and (7.12) it is clear that if we want to make the fracture toughness as large as possible we should do it using continuous fibres of uniform strength, rather than arranging for the fibre pull-out work to be large.

7.6. Oblique Fibres

Cracks propagate obliquely to fibres when the fibres are randomly dispersed in the matrix, and can also do so when the fibres are aligned, if the crack is suitably constrained.

When fibres cross cracks obliquely, extra energy-absorbing mechanisms can be identified, in both fibres and matrix. Figure 7.7 shows a ductile fibre which has been fractured by a crack passing obliquely. The fibre has suffered plastic bending, and it has plastically deformed the matrix near the crack face.

Brittle fibres, on the other hand, can be weakened by the flexure that occurs as the crack faces separate, thus giving a reduction in fracture work.

In this section we will consider aligned fibres, crossing a crack at an angle ϕ to the crack plane normal (Fig. 7.8). Brittle fibres (e.g. glass, boron, and carbon) will break at a reduced stress, σ_{fmax}, because of the bending near the crack plane, where

$$\sigma_{fmax} = \sigma_{fu}(1 - A \tan \phi). \tag{7.13}$$

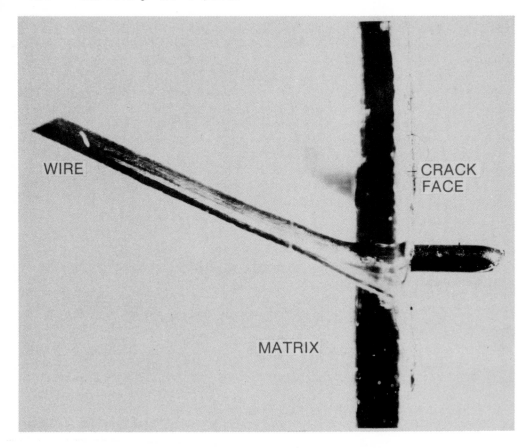

FIG. 7.7. Micrograph of a 0·5-mm steel wire in polycarbonate which crossed a crack obliquely, and was stressed to failure by opening of the crack.

Experiment has shown that $A \simeq 5.5\tau_{my}/\sigma_{fu}$. The critical aspect ratio for pull-out is thus also reduced:

$$s_{c\phi} = s_c(1 - A\tan\phi). \tag{7.14}$$

The plastic deformation of the matrix, as it is pushed aside during fibre flexure near the crack plane, introduces a factor $1 + 0.72\varepsilon_{fu}\tan^2\phi$ into the equation. Thus

$$\mathbf{G}_{fb\phi} = \mathbf{G}_{fb}(1 - A\tan\phi)^3(1 + 0.72\varepsilon_{fu}\tan^2\phi). \tag{7.15}$$

Fig. 7.9 shows a plot of $\mathbf{G}_{fb\phi}$ as a function of ϕ.

For shorter fibres, with $s > s_{c\phi}$, there is some pull-out and some stress redistribution. If we sum the appropriate amounts of work, as for equation (7.9) we have,

$$\mathbf{G}_{i\phi} = \frac{1}{6}V_f d\tau_i s_{c\phi}^3 \left\{ \frac{1}{s} + \frac{8\tau_i}{E_f}(1 + 0.72\varepsilon_{fu}\tan^2\phi)\left(1 - \frac{s_{c\phi}}{s}\right) \right\}. \tag{7.16}$$

Ductile fibres are not weakened by the flexure at the crack face, so that $\sigma_{fm} = \sigma_{fu}$. Thus

$$\mathbf{G}_{fb\phi} = \mathbf{G}_{fb}(1 + 0.72\varepsilon_{fu}\tan^2\phi) \tag{7.17}$$

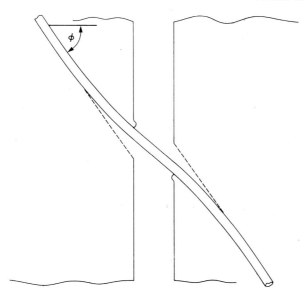

Fig. 7.8. Fibre crossing a crack diagonally.

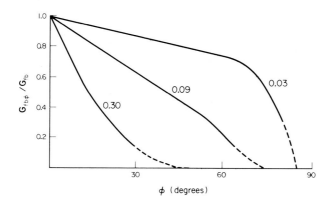

Fig. 7.9. Effect on toughness of angle between fibres and crack plane normal for $\sigma_{my}/\sigma_{fu} = 0.3$, 0.09, and 0.03. The dashed lines indicate regions where approximations are less good.

for continuous ductile fibres, and for short fibres, instead of equation (7.16), we have,

$$\mathbf{G}_{i\phi} = \frac{1}{6} V_f d\tau_i s_c^3 \left\{ \frac{1}{s} + \frac{8\tau_i}{E_f}(1 + 0.72\varepsilon_{fu}\tan^2\phi)\left(1 - \frac{s_c}{s}\right) \right\} \tag{7.18}$$

For short fibres, $s < s_{c\phi}$, we use $\mathbf{G}_{fp}$.

When ductile fibres are pulled out, extra work is done, when they are not normal to the crack plane, because of the plastic deformation of the fibre as it bends around the corner where it emerges at the crack face. The shearing of a single fibre with embedded length l absorbs an amount of energy U_{fs}, where

$$U_{fs} = \tfrac{1}{8}\pi r^2 \sigma_{fy} l \tan\phi$$

for $\tan\phi > 2$, where σ_{fy} is the yield stress of the fibre. For $\tan\phi > 2$ the fibres break because they are unable to withstand the shearing forces.

The number of fibres with an embedded length between l and $l+dl$ is $dN = Ndl/2L\cos\phi$ per unit area, if N is the number of fibres per unit area normal to the fibre direction. These fibres contribute to the work of fracture an amount of internal shear work given by

$$U_{fs}dN = \frac{V_f\sigma_{fy}\sin\phi}{16L\cos^2\phi}ldl \qquad (7.19)$$

and for long fibres $(s > s_c)\mathbf{G}_{fs}$ is twice this, integrated for all l between 0 and l_c:

$$\mathbf{G}_{fs} = \frac{V_f\sigma_{fy}s_c^2 d\sin\phi}{32s\cos^2\phi}. \qquad (7.20)$$

For short fibres, with $s < s_c$, we integrate equation (7.19) for all l between 0 and L and multiply by two. Thus

$$\mathbf{G}_{fs} = \frac{V_f\sigma_{fy}sd\sin\phi}{32\cos^2\phi}. \qquad (7.21)$$

7.7. Random Fibres

We will consider the planar random case. The work of fracture is obtained by integrating the appropriate work term, at angle ϕ, for all ϕ from zero up to the maximum value, ϕ_c, at which there is still a contribution to the fracture work. We consider one quadrant, so we divide the result by the range of possible orientations in the quadrant, i.e. $\pi/2$. Thus

$$\mathbf{G} = \frac{2}{\pi}\int_0^{\phi_c}\mathbf{G}(\phi)d\phi$$

The integration of the trigonometric functions in equations (7.15) and (7.16) presents some difficulties, and the solutions available are very approximate.
For brittle fibres we have

$$\mathbf{G}_{ir} \simeq \frac{1}{6}V_f d\tau_i s_{cr}^3 \left\{\frac{1}{s} + \frac{8\tau_i}{E_f}(1 - s_{cr}/s)\right\} \qquad (7.22)$$

where

$$s_{cr} = 2s_c\{\phi_c - A\ln(\cos\phi_c)\}/\pi \qquad (7.23)$$

and

$$\tan\phi_c = 1/2.4A. \qquad (7.24)$$

For ductile fibres

$$\mathbf{G}_{ir} = 0.12V_f d\tau_i s_c^3 \left\{\frac{1}{s} + \frac{8\tau_i}{E_f}(1 + 0.58\sigma_{fu}/E_f)(1 - s_c/s)\right\}. \qquad (7.25)$$

The work of fibre shearing can more easily be calculated. For long fibres $(s > s_c)$ we integrate equation (7.20) with respect to ϕ and obtain,

$$\mathbf{G}_{fsr} = 0.024V_f d\sigma_{fy}s_c^2/s \qquad (7.26)$$

while for short fibres we integrate equation (7.21) with respect to ϕ:

$$\mathbf{G}_{fsr} = 0.024 V_f d\sigma_{fy} s. \tag{7.27}$$

For $s < s_{cr}$ there is also the pull-out work $2\phi_c \mathbf{G}_{fp}/\pi$.

Fibres having high yield stresses, and aspect ratios close to s_c can contribute very greatly to the work of fracture. For example, steel with $\sigma_{fy} = 3\,\text{GPa}$, $d = 0.1$ mm and $V_f = 0.5$, in a polymer giving $\tau_i = 6\,\text{Mpa}$ contributes $\mathbf{G}_{fsr} \simeq 0.9\,\text{MJ m}^{-2}$ when the fibres have the critical aspect ratio. ($s_c \simeq 250$, if we assume $\sigma_{fu} \simeq \sigma_{fy}$. The COD is 12.5 mm.) This mechanism is employed in some reinforced ceramics, for example $\text{Mo–Al}_2\text{O}_3$.

7.8. Work of Fibre Fracture

This can be neglected with brittle fibres, such as boron, carbon, and glass. It is important with Kevlar and steel, however. For $s > s_c$ the fraction broken is $(1 - s_c/s)$, so if each fibre has a work of fracture (per unit area) of $\mathbf{G}_f$, the total work, $\mathbf{G}_{fd}$, is

$$\mathbf{G}_{fd} = V_f \mathbf{G}_f (1 - s_c/s). \tag{7.28}$$

The cellulose fibres in wood have a very high work of fracture. This is because they are hollow, and the walls have a spiral structure. Splitting and crumpling of the walls occurs when the fibres break, as shown in Fig. 7.10, and this involves a great deal of work. This observation raises the exciting possibility of designing fibres with high works of fracture which could be used to make tough composites.

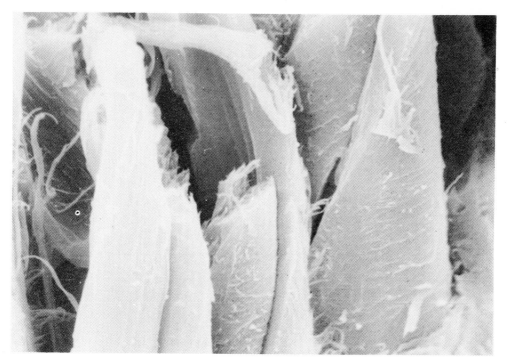

FIG. 7.10. Fractured wood showing splitting and other damage of cellulose fibres. (Courtesy of Jeronimidis, G., Reading University.)

7.9. Debonding and Splitting

Aligned fibre-polymers and ceramics can be split parallel to the fibres very easily. Wood presents a good example of this type of behaviour, as can be seen from the data in Table 1.2.

Near the crack tip, in addition to the very high stress normal to the crack plane, there is a high stress in the crack propagation direction, as shown in Fig. 2.11. This stress can cause splitting in aligned fibre composites, when the fibres are normal to the crack plane, as shown in Fig. 7.11. The splitting reduces the stress concentration at the crack tip, and strongly inhibits crack propagation. The process is difficult to quantify, and little effort has been made to take advantage of it because material which splits easily has poor shear properties.

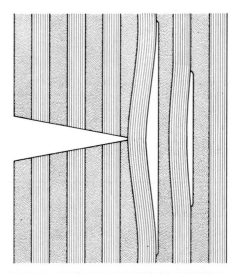

FIG. 7.11. Splitting caused by tensile stresses in crack propagation direction. Applied stress is normal to crack plane.

Fibre debonding can give some enhancement of toughness, even if fibre failure takes place in the crack plane. If the work of debonding (per unit area) is G_s, then the amount of work done in debonding a fibre up to a distance l from the crack face is $2\pi r l G_s$, at each surface. The contribution this makes to the work of fracture is G_{db} where

$$G_{db} = 4V_f l G_s/r \qquad (7.29)$$

G_s is usually quite small, $\approx 10\ \text{Jm}^{-2}$ in reinforced plastics and ceramics. For a moderate debonded length of 50 fibre diameters, in a composite with $V_f = 0.5$, G_{db} comes to about $0.5\ \text{kJ m}^{-2}$. This effect is usually unimportant, since the work involved is small. In any case, if we increase G_s very greatly, debonding is prevented.

7.10 Synergistic Effects

In certain circumstances fibres and matrix can work together to enhance a fracture property of one or the other. Two examples of this will be described.

Ductile fibres contribute substantially to the composite work of fracture by virtue of their high internal fracture work. This can be enhanced by the presence of the matrix, which constrains the fibres, inhibits necking, and increases the length of fibre which has undergone large plastic strains. Thus G_f is increased and is given by

$$G_f^* = \sigma_{fu} U_f d/2\tau_i. \tag{7.30}$$

The fibres deform over a length of about $s_c d$ (the critical length). U_f is the work done in deforming unit volume of the fibre material to its ultimate tensile strain.

A work of fracture of $50\,\mathrm{kJ\,m^{-2}}$ was obtained with aligned stainless steel–aluminium as a result of this process. This mechanism can also operate effectively in polymers and ceramics.

Fine fibres can increase the fracture strain of brittle matrices. V_f must be large before this happens, as shown in Fig. 7.12. The process is governed by the equation

$$\varepsilon_m^3 = \frac{12\tau_i G_m E_f V_f^2}{V_m E_m^2 d(V_f E_f + V_m E_m)} \tag{7.31}$$

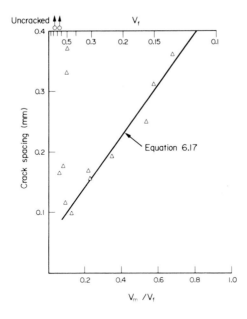

Fig. 7.12. Suppression of multiple fracture at high V_f for steel-epoxy at 77k. (After Cooper, G. A. and Sillwood, J. M. (1972) *J. Mater. Sci.*, **7**, 325.)

When ε_m (given by equation (7.31)) is greater than σ_{mu}/E_m matrix cracks are not generated until $\varepsilon_1 = \varepsilon_m$, for aligned fibre composites. Note that the effect is proportional to $d^{-1/3}$. In Fig. 7.12 the crack spacing is a linear function of V_m/V_f down to about $V_f = 0.4$, as would be expected from equation (6.17). For $V_f > 0.4$ the cracking is suppressed.

7.11. Cohibitive Effects

Sometimes the matrix and fibres interact unfavourably to reduce a fracture property of one or the other. The most well-understood cohibitive effect is the reduction in thickness of the worked zone in the matrix at the crack surfaces. This reduces G_m.

The plastically deformed zone in ductile metals can be 1 cm or more thick at each crack face. When fibres are present it is reduced to thickness t, where

$$t = \frac{V_m \sigma_{mu} d}{4\tau_i V_f}. \tag{7.32}$$

For a well-made reinforced metal $\tau_i \simeq \sigma_{mu}/2$. Thus if $V_f = 0.5$, $t = d/2$, or less than 6 μm for carbon or glass reinforcement. Even with boron reinforcement $t < 50$ μm. This means that the toughness contribution from even the most ductile matrix is reduced to a very small fraction of $V_m G_m$, which we shall express as $V_m G_m^*$.

7.12. Total Fracture Work

Instead of G_1 being given by a Rule of Mixtures expression we have

$$G_1 > V_f G_f^* + V_m G_m^* \tag{7.33}$$

for continuous aligned fibres, often with $G_f^* \gg G_f$ and $G_m^* \ll G_m$. G_1 can be very much greater than the right hand side of inequality 7.33, since the pull-out work and stress redistribution work can be very large. Estimation of G_1 must include all the contributions that are appropriate to the situation. In the case of short fibres the G_f^* term is reduced, and disappears altogether when $s < s_c$.

Further Reading

COOPER, G. A. and PIGGOTT, M. R. (1977), *Fracture* 1, 557.
Fracture Mechanics of Composites (1975) ASTM STP593 (ASTM, Philadelphia).
Composite Materials, Testing and Design (1977) ASTM STP617 (ASTM, Philadelphia).
CORTEN, H. T. (1972) *Fracture* (Academic Press, London) 7 Chapter 9

8
Reinforcement with Platelets

THIN sheets of such materials as silicon carbide and mica can be quite strong and have a higher Young's modulus than glass. In addition, mica is very cheap. Reinforcement with mica platelets seems quite an attractive alternative to glass fibre reinforcement where some strength and toughness can be sacrificed in favour of lower cost and improved modulus. The benefit in improved modulus is greater than indicated simply by the difference between the moduli because aligned platelets give two-dimensional stiffening, while aligned fibres only bestow substantial stiffness in the fibre directions.

In this chapter we use the simplified slip theory described in Section 4.1 to determine the role of aspect ratio. Aspect ratio must be considered very carefully in platelet reinforcement, since the possibility of continuous platelet reinforcement is restricted to very few special cases.

8.1. Square Platelets

We will consider platelets aligned with an edge parallel to the applied stress. The aspect ratio of the platelet is the ratio of edge length to thickness. The treatment here can be adapted for oblong platelets similarly aligned, so long as we put the correct aspect ratio into the equations: this is the length of the side parallel to the stress divided by the platelet thickness. For the treatment to be valid the platelet thickness must be constant.

For platelets aligned in other ways we must make other assumptions, as described in Section 8.2 for round platelets.

8.1.1. Stress Transfer

There are slip regions near at the ends of the platelet as shown in Fig. 8.1. We neglect stress transfer at the edges parallel to the applied stress. Thus, in an element of length dx near the end, the shear stress operates over an area $2wdx$, where w is the width of the platelet. The force exerted by the platelet stress, σ_p, is $\sigma_p tw$ where t is the platelet thickness. These forces must be equal. Thus

$$\frac{d\sigma_p}{dx} = -\frac{2\tau_i}{t}. \tag{8.1}$$

The plateau stress, i.e. the stress in the centre region, σ_{pp}, is

$$\sigma_{pp} = 2\tau_i mL/t.$$

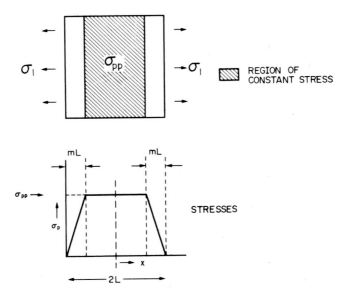

FIG. 8.1. Stresses in a single embedded square platelet.

The aspect ratio, s, is

$$s = 2L/t \qquad (8.2)$$

so our equation for σ_{pp} becomes

$$\sigma_{pp} = \tau_i ms. \qquad (8.3)$$

As in the slip theory of fibre reinforcement we equate the fibre strain in the unslipped region with the matrix strain and composite strain. Thus

$$\varepsilon_1 = \sigma_{pp}/E_p \qquad (8.4)$$

where E_p is the platelet modulus. Consequently

$$m = \frac{E_p \varepsilon_1}{\tau_i s} \qquad (8.5)$$

We cannot estimate the composite stress until we have taken the platelet packing arrangement into account. With fibres, the load that is shed by one fibre, near its end, is transferred to five or six nearest neighbours. The resulting stress concentration is usually not significant. With platelets, however, at least a 50% increase in stress can be expected in adjacent platelets near the end of one platelet. This cannot be ignored.

8.1.2. Symmetrical Packing

Consider the arrangement shown in Fig. 8.2a. Platelets having aspect ratios that are sufficiently large will have the stress distribution shown in Fig. 8.2c. The maximum platelet

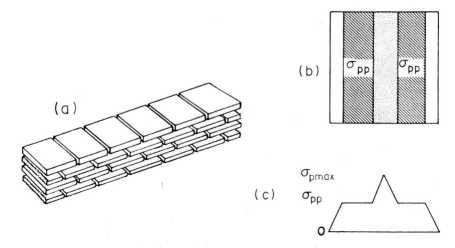

FIG. 8.2. (a) Symmetrical packing arrangement for square platelets, (b) stress regions in platelets, and (c) stress distribution in platelets.

stress, σ_{pmax}, is twice the plateau stress, i.e.

$$\sigma_{pmax} = 2\sigma_{pp} \tag{8.6}$$

because of the absence of platelets in alternate layers in the cross-section where $\sigma_p = \sigma_{pmax}$. It can be readily seen that the average platelet stress, $\bar{\sigma}_p$ is

$$\bar{\sigma}_p = \sigma_{pp} \tag{8.7}$$

since the area under the stress spike at the platelet centre is equal to the area under the sloping regions of the stress distribution near the platelet ends. (See Fig. 8.2c.)

The platelets and matrix will obey the Rule of Averages (equation (4.5)) which here takes the form

$$\sigma_1 = V_p\bar{\sigma}_p + V_m\bar{\sigma}_m. \tag{8.8}$$

We assume $\bar{\sigma}_m = E_m\varepsilon_1$ and using equations (8.7) and (8.3) for $\bar{\sigma}_p$ and σ_{pp} and equation (8.5) for m, the Rule of Averages becomes

$$\sigma_1 = (V_pE_p + V_mE_m)\varepsilon_1 \tag{8.9}$$

for $\sigma_m < \sigma_{my}$. Since $E_1 = \sigma_1/\varepsilon_1$ we have

$$E_1 = V_pE_p + V_mE_m. \tag{8.10}$$

Thus the platelet composite obeys the Rule of Mixtures for modulus at all platelet aspect ratios. This result has been confirmed over a wide range of aspect ratios for steel disc-polycarbonate.

The composite strength depends on whether or not the platelet aspect ratio is greater than the critical aspect ratio. As with fibres, we define the critical aspect ratio as the aspect ratio just big enough for the matrix to transfer the breaking stress to the platelet. We can calculate s_c using equation (8.3) with $m = \frac{1}{2}$ and $\sigma_{pp} = \sigma_{pu}/2$. Thus

$$s_c = \sigma_{pu}/\tau_i. \tag{8.11}$$

When $s > s_c$, $\sigma_{pmax} = \sigma_{pu}$ at the composite breaking-stress, and $\bar{\sigma}_p = \sigma_{pp} = \sigma_{pu}/2$. Equation (8.8) then gives the composite breaking stress, σ_{1u} as

$$\sigma_{1u} = \tfrac{1}{2} V_p \sigma_{pu} + V_m \bar{\sigma}_m \qquad (8.12)$$

$\bar{\sigma}_m = \sigma_{pu} E_m / 2 E_p$, or $\bar{\sigma}_m = \sigma_{my}$ if $\varepsilon_{pu} > 2\varepsilon_{my}$ since $\varepsilon_{1u} = \tfrac{1}{2}\varepsilon_{pu}$. Equation (8.12) shows that we can achieve only half the Rule of Mixtures strength when platelets are packed in this fashion. We will define a strength efficiency factor, χ_p, by the equation

$$\sigma_{1u} = \chi_p V_p \sigma_{pu} + V_m \bar{\sigma}_m. \qquad (8.13)$$

For this packing arrangement, and for $s > s_c$, $\chi_p = \tfrac{1}{2}$.

When $s < s_c$ we get failure by slip rather than by platelet fracture. The matrix will start to slip past the platelets when $m = \tfrac{1}{2}$, i.e. when

$$\sigma_{pp} = \tau_i s/2. \qquad (8.14)$$

The strain at the onset of this gross slip, ε_{1p},

$$\varepsilon_{1p} = \tau_i s/2E_p. \qquad (8.15)$$

When $\varepsilon_1 > \varepsilon_{1p}$ the platelet stress is constant, i.e. independent of ε_1, and the Rule of Averages gives

$$\sigma_1 = V_p \tau_i s/2 + V_m E_m \varepsilon_1 \qquad (8.16)$$

for $\varepsilon < \varepsilon_{my}$. When $\varepsilon_1 > \varepsilon_{my}$

$$\sigma_1 = V_p \tau_i s/2 + V_m \sigma_{my}. \qquad (8.17)$$

A non-work hardening matrix will have $\sigma_{my} = \sigma_{mu}$, so the composite should break when $\sigma_1 = \sigma_{1u}$ where

$$\sigma_{1u} = V_p \tau_i s/2 + V_m \sigma_{mu}. \qquad (8.18)$$

Comparing this with equation (8.13) we find that

$$\chi_p = \tau_i s/2\sigma_{pu}. \qquad (8.19)$$

We can now draw stress–strain curves, and these are shown schematically in Fig. 8.3.

Note that the same relations hold for σ_2 and ε_2, and $\sigma_{2u} = \sigma_{1u}$. The material is approximately transversely isotropic so long as the packing arrangements are the same in both directions.

8.1.3. Non-symmetrical Packing

Figure 8.4a shows a representative non-symmetrical arrangement, and Fig. 8.4c shows the corresponding stress distribution in the platelets at low applied stress.

We recognize three cases: (1) $s > 3s_c/2$, (2) $s_c < s < 3s_c/2$ and (3) $s < s_c$. s_c is defined as previously (equation (8.11)).

(1) $s > 3s_c/2$. In this case the stress distribution shown in Fig. 8.4c applies up to the breaking-stress. Geometrical considerations (Fig. 8.4c) lead to

$$\bar{\sigma}_p = \sigma_{pp}(1 - m/4). \qquad (8.20)$$

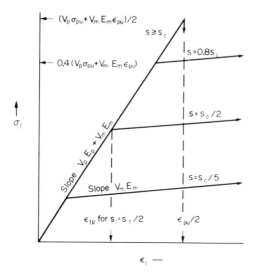

<figure>FIG. 8.3. Schematic stress–strain curve for symmetrically packed square platelet reinforced polymer ($\varepsilon_{my} > \varepsilon_{pu}/2$).</figure>

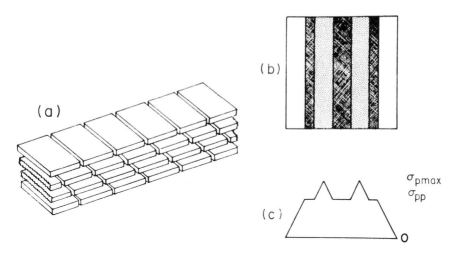

FIG. 8.4. (a) Non-symmetrical packing arrangement for square platelets, (b) stress regions in platelets, and (c) stress distribution in platelets.

We use the Rule of Averages to calculate the applied stress, together with equations (8.4), (8.5), and (8.20):

$$\sigma_1 = (V_p E_p + V_m E_m)\varepsilon_1 - \frac{V_p E_p^2 \varepsilon_1^2}{4\tau_i s} \tag{8.21}$$

for $\varepsilon_1 < \varepsilon_{my}$. This equation indicates an aspect ratio effect, albeit a small one, i.e. the stress–strain trajectory is slightly curved for aspect ratios near the lower limit.

The platelets break when $3\sigma_{pp}/2 = \sigma_{pu}$. Thus the composite breaking strain is

$$\varepsilon_{1u} = 2\sigma_{pu}/3E_p \qquad (8.22)$$

and the corresponding stress is obtained from equation (8.21) with $\varepsilon_1 = \varepsilon_{1u}$:

$$\sigma_{1u} = \frac{2}{3}V_p\sigma_{pu}(1 - s_c/6s) + 2V_mE_m\sigma_{pu}/3E_p \qquad (8.23)$$

for $\varepsilon_{1u} < \varepsilon_{my}$. For $\varepsilon_{1u} > \varepsilon_{my}$ replace the last term by $V_m\sigma_{my}$ for non-work hardening matrices.

Comparison with equation (8.13) gives

$$\chi_p = \frac{2}{3}(1 - s_c/6s). \qquad (8.24)$$

Thus the strength efficiency factor now has a maximum value of $2/3$.

(2) $s_c < s < 3s_c/2$. In this case the stress–strain trajectory has two regions. The stress distribution shown in Fig. 8.4c applies until $m = \frac{1}{2}$. The strain is then ε_{1p} (equation (8.15)), and for $0 < \varepsilon_1 < \varepsilon_{1p}$ equation (8.21) is obeyed.

When $m > \frac{1}{2}$ the stress plateau in Fig. 8.4c disappears and the stress distribution is a twin peak, separated by a V-shaped depression. At this stage

$$\bar{\sigma}_p \simeq \sigma_{pp}\left(3 - \frac{9m}{4} - \frac{1}{2m}\right)$$

and the same analysis used to develop equation (8.21) gives

$$\sigma_1 \simeq V_p\left\{E_p\varepsilon_1\left(3 - \frac{9E_p\varepsilon_1}{4\tau_i s}\right) - \tfrac{1}{2}\tau_i s\right\} + V_mE_m\varepsilon_1 \qquad (8.25)$$

for $\varepsilon_1 < \varepsilon_{my}$. The composite fails at a stress

$$\sigma_{1u} \simeq 2V_p\sigma_{pu}(1 - s_c/2s) - \tfrac{1}{2}V_p\tau_i s + V_mE_m\varepsilon_{1u} \qquad (8.26)$$

Comparison with equation (8.13) gives

$$\chi_p \simeq 2(1 - s_c/2s) - \tau_i s/2\sigma_{pu}. \qquad (8.27)$$

(3) $s > s_c$. In this case the stress–strain trajectory has three regions. At low strains ($m < \frac{1}{2}$) equation (8.21) is obeyed, and at higher strains ($\frac{1}{2} < m < 2/3$) equation (8.26) is obeyed. When $m = 2/3$, $\varepsilon_1 = \varepsilon'_{1p}$ where

$$\varepsilon'_{1p} = \frac{2\tau_i s}{3E_p}. \qquad (8.28)$$

For strains greater than ε'_{1p} the stress is given by equation (8.16) or (8.17) and the platelet stress is independent of strain. χ_p is given by equation (8.19).

Figure 8.5 is a schematic drawing of the stress–strain trajectories for various aspect ratios. In all three cases, when ε_1 reaches ε_{my}, $\varepsilon_1 E_m$ should be replaced by σ_{my}. Figure 8.6 shows χ_p plotted as a function of s/s_c. The results for symmetric packing are also shown in the figure.

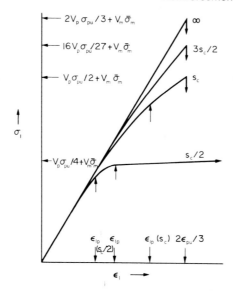

FIG. 8.5. Schematic stress–strain curves for square platelet-polymers, packed as in Fig. 8.4.

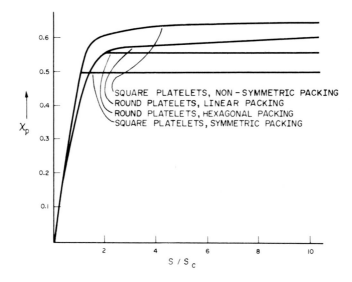

FIG. 8.6. Theoretical prediction for strengthening efficiency factor, χ_p, as a function of aspect ratio.

8.2. Round Platelets

Figure 8.7 shows the stress distribution assumed for a single round platelet. We have treated the platelet as though we can divide it up into infinitely thin strips, parallel to the applied stress, and not quite touching each other. We are therefore neglecting stress

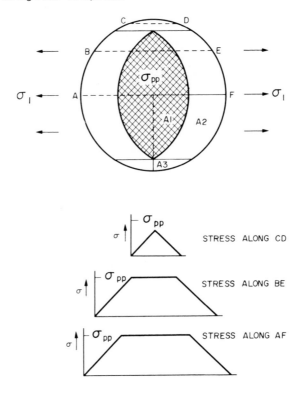

FIG. 8.7. Stresses in single embedded round platelet assuming isobars are parallel to platelet edge.

transfer within the platelet, which may be justified only when the platelet is very thin.

The aspect ratio of the platelet is the ratio of diameter to thickness. Stress transfer is governed by equations (8.1) and (8.3), and the composite strain is given by equation (8.4), so that m is given by equation (8.5).

Figure 8.8a shows a symmetrical linear packing arrangement, and the stress distribution when the platelets are about to break is shown in Fig. 8.8b. Experiments with steel platelet–polycarbonate show that the platelets break across their diameters (Fig. 8.9), as expected from the stress distribution.

Another packing arrangement has also been analysed. In this the fibres are packed hexagonally. The corresponding stress distribution at the failure point is shown in Fig. 8.10.

Values for χ_p for these packing arrangements are plotted vs. s/s_c in Fig. 8.6. Note that there is very little aspect ratio effect for $s > s_c$. This has been confirmed, for the linear arrangement, by experiments with steel platelet–polycarbonate. The maximum value of χ_p is little more than one-half (see Fig. 8.11). Thus platelets do not have much advantage over fibres on account of their two-dimensional reinforcement. Cross-ply fibre laminates are just as efficient, so long as care is taken to prevent matrix splitting parallel to the fibres. The platelet composites do, however, give more efficient stiffening than cross-ply fibre laminates.

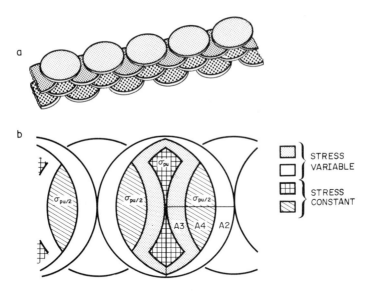

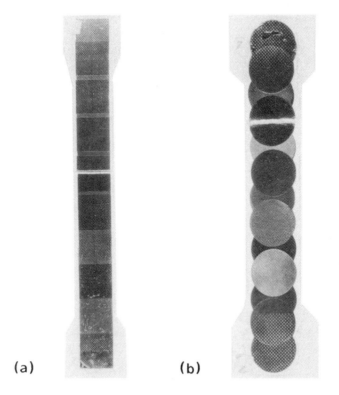

FIG. 8.8. Linear packing arrangement for circular platelets, and corresponding stress regions for platelets about to break.

FIG. 8.9. Specimens of symmetrically packed steel reinforced polycarbonate after testing to failure in tension. (a) Square and (b) round discs.

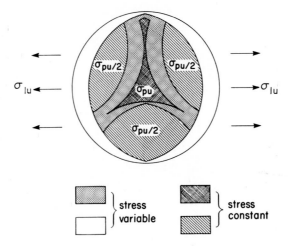

FIG. 8.10. Schematic drawing of stresses in hexagonally packed circular platelets.

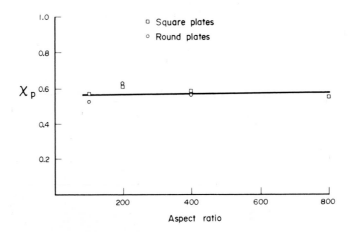

FIG. 8.11. Experimental results for strength efficiency factor, χ_p.

8.3. Toughness

Little work has been done on the toughness of platelet reinforced materials, though soft metal-hard metal laminates have been examined in some detail. Two configurations have been found to be effective, and are shown in Fig. 8.12. When the crack front is normal to the laminae (Fig. 8.12a) the crack propagation in the hard laminae is inhibited by the soft layers. When the crack front is parallel to the laminae (Fig. 8.12b) the laminae tend to separate just ahead of the crack, and this has the effect of reducing the stress concentration and diverting the crack. In both cases considerable improvements in toughness are obtained.

Such processes are also at work when aligned platelets are surrounded by a soft matrix. The toughening effect is hard to quantify, however.

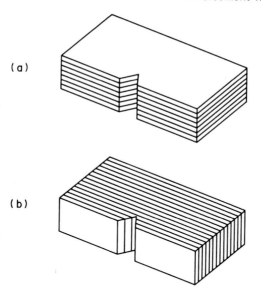

(a)

(b)

FIG. 8.12. (a) Crack divider and (b) crack arrester.

The laminae will contribute pull-out work, and stress relaxation work (for $s > s_c$) whether they are aligned as in Fig. 8.12a or Fig. 8.12b, or anywhere between. We will consider these contributions to toughness for square platelets. (The same treatment is appropriate for platelets with other shapes, but the resulting equations are rather cumbersome.)

8.3.1. Pull-out

Consider aligned square platelets that are normal to the crack plane with edges that are parallel to it (Fig. 8.13). The pull-out work U_{pp}, for a single platelet with an embedded length l is

$$U_{pp} = \int_0^l 2wx\tau_i dx \tag{8.29}$$

(Compare this with equation (7.2) for a single fibre). Following the treatment for fibres, we calculate the total work for all platelets, $\frac{1}{2}\mathbf{G}_{pp}$, on each side of the crack. This is

$$\tfrac{1}{2}\mathbf{G}_{pp} = \frac{N}{2L} \int_0^L U_{pp} dl$$

where L is the platelet length and N is the number of platelets per unit area in the cross-section being fractured. Since $V_p = Ntw$ this comes to

$$\mathbf{G}_{pp} = V_p t \tau_i s^2 / 12 \tag{8.30}$$

This is one-half the corresponding value for the fibre case (equation (7.3)).

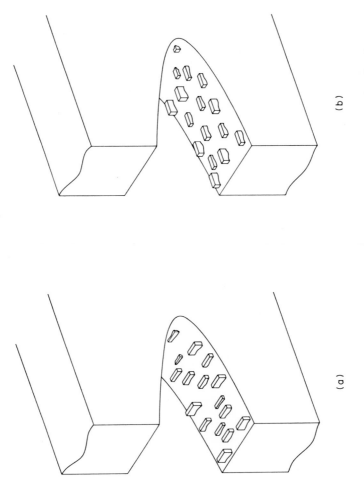

FIG. 8.13. Platelets oriented (a) across and (b) parallel to the crack propagation direction.

The maximum value of $\mathbf{G}_{pp}$ is for $s = s_c$. In this case

$$G_{pp\,max} = V_p t \sigma_{pu}^2/12\tau_i \qquad (8.31)$$

which is twice the corresponding value for the fibre case.

For a mica-epoxy, with $\sigma_{pu} = 0.85$ GPa, $\tau_i = 8$ MPa, $t = 1\ \mu$m and $V_p = 0.8$, $\mathbf{G}_{pp\,max}$ comes to about 6 kJ m^{-2}. This is very much less than the corresponding value for carbon-epoxy (150/kJ m^{-2}), partly because of the lower strength of the mica, and partly because of the very small thickness of the mica.

In practice, mica used for reinforcement has $s \leqslant 100$. For an average value of about 50 for s, pull-out occurs. We then use equation (8.31) with the same values for t, τ_i, and V_p as before, with $s = 50$. Thus $\mathbf{G}_{pp}$ comes to about 1.6 kJ m^{-2}.

8.3.2. Stress Relaxation

Consider continuous platelets. We follow the treatment of Section 7.2. The stored energy for a short length of platelet, dU_{pb}, at distance $x(< ts_c/2)$ from the crack face is

$$dU_{pb} = \frac{wt\,\sigma_p^2 dx}{2E_p}$$

where the platelet stress, σ_p, is

$$\sigma_p = 2\tau_i(l_c - x)/t$$

where l_c is the length required to transfer the platelet breaking-stress. Thus U_{pb} is obtained by integration

$$U_{pb} = \int_0^{l_c} \frac{2\tau^2 (l_c - x)^2 w dx}{E_p t}$$

The work done by the element of platelet sliding against the matrix, U_{mp}, is the same as this, so that $\frac{1}{2}\mathbf{G}_{pb} = N(U_{pb} + U_{mp})$. Thus

$$G_{pb} = \frac{V_p t \sigma_{pu}^3}{3E_p \tau_i} \qquad (8.32)$$

since $l_c = \sigma_{pu} t/2\tau_i$.

The mica composite which gave $\mathbf{G}_{pp\,max} = 6\ \text{kJ}\,\text{m}^{-2}$ gives $\mathbf{G}_{pb} = 90\ \text{Jm}^{-2}$ since $E_p = 226$ GPa. As with fibres $\mathbf{G}_{pb}$ is very much less than $\mathbf{G}_{pp\,max}$. Note that the ratio $\mathbf{G}_{pp\,max}/\mathbf{G}_{pb} = E_p/4\sigma_{pu}$ is the same as the analogous ratio for fibre reinforcement.

8.3.3. Platelets with $s > s_c$

We carry out the same integrations as with fibres to calculate the work, $\mathbf{G}_i$. Thus

$$\tfrac{1}{2}\mathbf{G}_i = \frac{N}{2L} \int_0^{l_c} U_{pp} dl + \frac{1}{2L} \int_{l_c}^{2L - l_c} 2U_{pb} dl$$

so that

$$G_i = \frac{V_p t \sigma_{pu}^3}{3\tau_i} \left\{ \frac{1}{4\tau_i s} + \frac{1}{E_f}\left(1 - \frac{\sigma_{pu}}{\tau_i s}\right) \right\}. \qquad (8.33)$$

This should be compared with equation (7.9) for fibres. It gives exactly the same plot of G_i as a function of aspect ratio as shown in Fig. 7.4.

8.3.4. Cohibitive effect

The stress concentrations in the platelets that arise from the loads shed by adjacent platelets near their boundaries causes the composite to fail when $\varepsilon_1 < \varepsilon_{pu}$. Figure 8.14 shows the stress–strain curve obtained in experiments with steel platelet–polycarbonate,

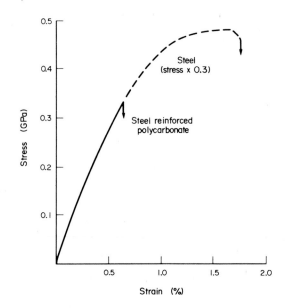

FIG. 8.14. Stress–strain curve for steel-polycarbonate, and steel with stress × 0.3.

with the curve for the steel superposed. (The superposition was effected by multplying the stress for the steel by 0.3.) The failure strain of the steel in the polycarbonate is less than half its failure strain when not embedded. The plastic region of the stress–strain curve is almost completely suppressed, so that the platelets are prevented from contributing their work of fracture to that of the composite.

Further Reading

KATZ, H. S. and MILEWSKI, G. V. (eds.) (1978) *Handbook of Fillers and Reinforcements for Plastics* (Van Nostrand, Reinhold, New York), Sections 8 and 9.
GLAVINCHEVSKI, B. and PIGGOTT, M. R. (1973) *J. Mater. Sci.* **8**, 1373.

9

Reinforced Polymers

REINFORCED polymers have a history dating back to the early years of this century. The first United States patent for a structural composite was taken out in 1916; this was a tube made by hot pressing layers of fibrous material with a suitable binder.

Natural fibres, obtained from wood or flax, were used in early composites; they were prepared in the form of papers and fabrics, and embedded in phenolic resins by hot pressing. Strengths of 200 MPa were readily obtainable, with moduli of 20 GPa and a density of about $1.5\,Mg\,m^{-3}$. These composites were a considerable improvement on unreinforced plastics; the matrix alone has a tensile strength of 40 MPa, and a modulus of 3.5 GPA. The cellulose fibres, however, were affected by moisture, and in high humidity environments the composites swelled and became weaker.

The material which is popularly called "fibreglass" (a misnomer, since the word describes only one part of the composite) and which is used widely these days in boats, specialty cars, and a host of other structures, was developed originally by the military for radar domes on aircraft in 1941. It consists of glass-polyester and has greatly superior properties to cellulose-polymers.

Glass-polymers of a great variety of types are now widely used, and a number of other types of fibres are being used to reinforce polymers. This chapter will be concerned mainly with fibre reinforced composites which have high strength and modulus, and can perform a useful function in load bearing situations. Manufacturing methods will be considered first, then mechanical properties will be described.

9.1. Methods of Manufacture

The way in which a composite is made depends a great deal on the polymer used. Thus a brief discussion of polymer matrices will be necessary first, together with a discussion of factors affecting the transfer of loads from the fibre to the matrix at the interface.

9.1.1. The Matrix

Polymers are extremely high molecular weight materials with carbon normally being in a high proportion. The carbon atoms, which may number in the millions in a single molecule, are connected together to form long chains and extensive networks. Other elements (most notably silicon together with oxygen—the silicones) can also form long chains and networks, and other elements (e.g. oxygen, nitrogen, and sulphur) can be present in the carbon chains. Attached to the chains, which form a backbone for the

155

structure, are hydrogen atoms, organic groups, inorganic groups and radicals.

The diversity of possible compounds is immeasurable, so there will always be exceptions to any general observations. However, we will attempt to describe their properties in a general way.

Polymers are usually characterized by low moduli and strengths. This is due to the deformability of the complex networks, and the sliding that can take place between the long chains. Compared with the ductile metals, they are not very tough, and indeed some are very brittle, especially at temperatures below 0 C. Polymers are not resistant to heat. Nylon, for example, softens at less than 100 C. Few polymers can be taken above 200 C without serious loss of properties, due to softening, melting, and permanent changes such as chain scission. Polymers that can survive at all 300 C are very rare. However, the versatility and ease of handling of polymers often more than makes up for these disadvantages, and during the last 50 years polymers have taken over from metals in a wide range of applications where load-bearing properties are not an overriding consideration. In any case, low modulus is sometimes an advantage. Polyethylene garden hoses and rubber tyres are successful because of their low moduli.

There are two distinct classes of polymers: thermoplastics and thermosets. The thermoplastics (of which nylon is a good example) can be melted, and thus can easily be shaped after the chemical reactions have taken place which produce the backbone—the polymerization reaction. In the molten state they are viscous liquids which can easily be moulded. Thermosets on the other hand, start off as viscous liquids at room temperature and usually require heat to promote the polymerization reaction. The product formed is a three-dimensional network which cannot be melted. Epoxy resin is a good example of a thermoset. The fusibility of the thermoplastics is due to their having a long chain or branch chain structure, in contrast to the network structure of the thermosets. Thermosets cannot be easily shaped after polymerization, and so are polymerized in moulds having the final shape required. Special, partly cured thermosets, have been developed for use as matrices in composites. These are soft and easily mouldable, and are cured by heating, usually while in the mould.

Thermosets are usually used for high performance load-bearing composites (often referred to as advanced composites) because of the relative ease of producing the composite without damaging the fibres significantly, and without the need to chop the fibres into short lengths.

When thermosets or thermoplastics are used, shrinkage of the matrix occurs during the manufacture of the composite. With thermoplastics the shrinkage is due to the thermal contraction in cooling from the melting point and so can be calculated directly. With thermosets, the process is more complex. Some data on shrinkage is given in Table 9.1. The shrinkage may be important for the composite, since it can provide a compressive radial stress at the fibre surface which would assist in transferring the stress from the matrix to the fibre.

The radial stresses observed in model composites are much less than those expected if we assume that we can use the product of the shrinkage strain and the modulus to calculate the stress. With steel reinforced polycarbonate the stress observed more than 30 days after manufacture was 7 MPa (cf. 35 MPa) while with steel-epoxy it was 3 MPa (cf. 40 MPa). These stresses were measured in fibre pull-out tests, on stresseed single fibre "composites". Some results are shown in Fig. 9.1. The residual stress can be determined from the intercept with the applied stress axis. Due to the visoelastic behaviour of

TABLE 9.1. *Linear Cure Shrinkage of Selected Polymers*

(These are representative values only; considerable deviations will be observed in practice.)

Thermosets	Epoxy	Polyester	Phenolic
Shrinkage (%)	2	4	2
Thermoplastics	Polycarbonate	Polyethylene	Nylon
Shrinkage (%)	1.4	0.6	2

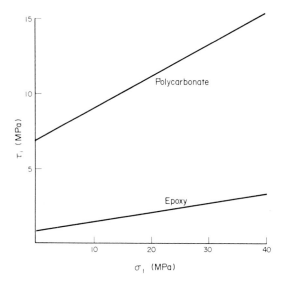

FIG. 9.1. Frictional stress vs. applied stress. Intercept with abscissa gives residual radial stress at fibre–matrix interface. (After Hadjis, N. and Piggott, M. R. (1977) *J. Mater. Sci.* **12**, 358.)

polymers it is quite possible that these stresses die away altogether after a sufficiently long period of time.

In sharp contrast to this, the individual layers can support stresses, and the layers within laminates have different expansion coefficients along the fibres and perpendicular to them. Thus, when there is a big difference between the temperature of manufacture, and that of actual use (this is normally the case), large residual stresses are observed, and can cause severe warping of unbalanced laminates (Fig. 9.2) and cracking of the matrix (Fig. 9.3).

9.1.2. *The Interface*

The state of the interface affects the efficiency of the transfer of loads from the fibres to the matrix. We showed in Section 5.2 that stress transfer occurs near the fibre ends, and

FIG. 9.2. Unsymmetric laminate warped due to thermal expansion differences. Kevlar fabric-epoxy. (Chamis, C. C. (1978) Proc. ICCM2, 221. Courtesy of the Metallurgical Soc., AIME.)

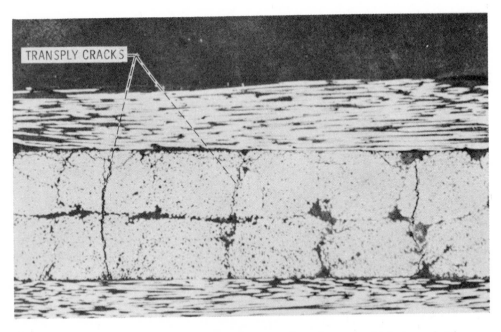

FIG. 9.3. Cracking-due to residual thermal stresses in carbon-epoxy laminate produced during manufacture. (Chamis, C. C. (1978) Proc. ICCM2, 221. Courtesy of the Metallurgical Soc., AIME.)

that, except when very low stresses are applied to the composite, the adhesion at the fibre–matrix interface fails near the fibre ends. In fact, adhesion only plays an indirect role in the transfer of stress. This point is usually overlooked, and a great deal of effort has been spent on improving the adhesion in the hope thereby of improving the efficiency of stress transfer.

Nevertheless adhesion is extremely important. Good adhesion is needed for two reasons: (a) it has a large effect on the properties of the composite in directions transverse to the fibres, when they are aligned, and on the shear properties of the composite, and (b) good adhesion reduces the rates of degradation of the composite in aggressive environments. A good example of the latter is the effect of water on glass reinforced plastics. Without good adhesion, degradation of mechanical properties occurs as water migrates to the fibre–matrix interface, and forms a thin film there. This action is promoted by the hydrophilic nature of the glass surface.

The method used to promote good adhesion depends a great deal on the fibre and matrix being used. In the case of glass, organic silane adhesion promoters are used. The silicone (SiO_2) group has a strong affinity for the glass while the organic part of the molecule has a strong affinity for the polymer. The molecule is thus expected to orient itself with the SiO_2 groups attached to the glass surface and the organic group embedded in the polymer. The organic group is chosen to give maximum compatibility with the particular matrix polymer to be used. The adhesion promoter is usually put onto the glass during manufacture of the fibres, as part of the protective layer for the glass surface. Thus, when ordering glass fibres for use as reinforcement it is usual to specify the polymer with which the fibres should be compatible.

Carbon is an inert material, and hence aggressive environments do not affect carbon fibres a great deal. The inertness of the carbon, however, makes the attainment of good adhesion extremely difficult. The solution normally adopted is to roughen the surface by direct oxidation, or acid attack. This gives a mechanical keying effect, and has a very beneficial effect on the coefficient of friction, which is important for stress transfer. Adhesion promoters are also used.

Kevlar and boron fibres adhere quite well to epoxy and polyester resins without treatment, but Kevlar should be dried before use.

The use of soft and yielding interfacial layers (e.g. rubbery polymers) has been demonstrated to be a practical way of increasing the toughness of composites. However, this does not appear to have had much application, as yet.

9.1.3. Premixes

The materials used to make fibre reinforced polymers are supplied in a number of different forms. The fibres are available as cloths, tapes, mats, etc. as described in Section 3.5, but it is often advantageous to combine the fibres with the resin prior to the production of an article. If a thermoset is used, it is combined with the fibres in a partly polymerized state, so that it can flow, when moulded, to take the desired shape. Once shaped, it is heated to complete the polymerization and make the shape permanent. Thermoplastics may also be combined with fibres prior to production. In this case the fibres are embedded in the molten polymer by liquid infiltration, and the solidified rods produced on cooling are chopped into short lengths. Heat is used during moulding to melt the polymer. Composites can also be moulded directly from mixtures of chopped fibres and powdered matrix.

The premixed materials have a number of important advantages. The quality of the product is reproducible, and the manufacture of components is simple, and can often be automated. The fibre content can be very high, yet they can be uniformly distributed. A wide range of these premixes is available, and their use takes the chemistry out of construction.

A widely used premixture is the sheet moulding compound. To make this, the fibres are cut into short lengths (e.g. 5 cm) and sandwiched between layers of partly cured thermoset supported by polyethylene films as shown in Fig. 9.4. The paste can contain other fillers, and pigments, etc. The mixture is then compacted by rolling, and wound up on a spool. It is then stored, so that the resin cures to the limited extent required, or the partial polymerization can be effected by heating, if the material is to be used without delay. The polyethylene surface films are removed just prior to pressing.

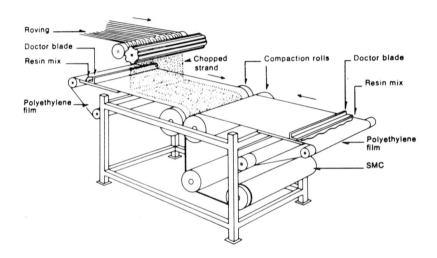

FIG. 9.4 Typical arrangement used for making sheet moulding compounds.
(Courtesy of PPG Industries.)

Another widely used premixture is the bulk moulding compound. This also contains short fibres (typically 3 to 30 mm) and also has a thermoset resin in a partly cured state, so that the mixture has a doughy consistency. It also contains additives of various types (e.g. pigments, etc.), and is supplied in bulk form as its name implies, or can be extended into rope-like form.

Continuous fibre–resin mixtures are usually called prepregs, and can be in the form of tapes or sheets, with woven or straight fibres. The resin used is a partly cured thermoset. Prepreg tapes are also made with aligned short fibres, and aligned whiskers.

Thermoplastic premixes are produced in the form of pellets. These contain short fibres (typically 3 mm), which have been embedded in the resin. Additives such as pigments may be included, and the pellets may contain the fibres at the concentration to be used for the product, or at higher concentration for subsequent dilution in the forming process.

A large number of different methods are used for the manufacture of the reinforced polymer product. The method chosen in any particular instance depends on the type and size of article being produced, the number of identical articles to be made, and the strength, stiffness, and other properties required of the material. The methods may be divided into two categories, those that require pressure and those that do not. In the next few sections some of the many moulding methods used for the manufacture of reinforced plastic articles will be described.

9.1.4. Hand Lay-up

The simplest method of making a composite is to lay the fibres onto a mould by hand, paint the resin on, and allow to cure. This requires very little capital investment, and though very labour-intensive, is still much used because of its great versatility.

The fibres can be in the form of random mat or cloth, and very large structures can easily be built if room temperature curing resins (e.g. epoxy or polyester) are used, and the resin can be sprayed rather than brushed on. The moulds do not need to be particularly strong, and can be made for example with balsa wood, and covered with plaster. More permanent moulds can be made with reinforced plastics.

Figure 9.5 illustrates this process. The gel coat is usually a pigmented resin layer to give good appearance, and may be sprayed on. When this is tacky the fibre mat or cloth is laid on manually, and more resin applied by pouring, brushing, or spraying. Next the layer is rolled, or a squeegee is used to ensure thorough impregnation and wetting of the fibres. Further layers are then added in the same way until the required thickness is obtained. The curing of the moulded part may be accelerated by heating.

Fig. 9.5. Resin is applied by brush in hand lay-up process. (Courtesy of Fibreglass, UK.)

9.1.5. Spray-up

In this method the fibres and resin are sprayed together onto the mould as shown in Fig. 9.6. The layers deposited are densified with rollers or squeegees as for hand lay-up. Gel coats are often used for good surface finish. Spray-up can be used to mould more complex shapes than hand lay-up. However, since the fibres have to be chopped in the spray gun, the composite produced cannot be as strong as hand laid-up composites with cloth or other continuous fibre forms.

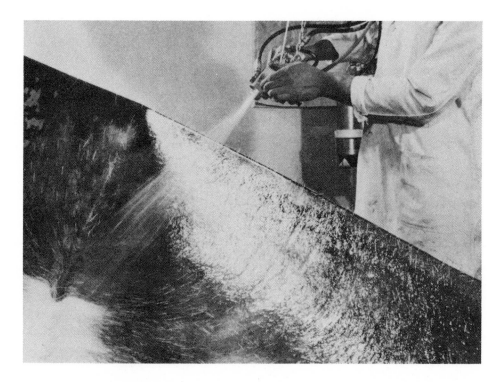

FIG. 9.6. Spray-up process for small boat making. (Courtesy of Fibreglass UK.)

Polyester or epoxy thermosetting resins are usually used. Very large parts are cured at room temperature, but with smaller parts the curing may be accelerated by bag moulding (Section 9.1.8). The moulds used do not need to be strong, and can be made with wood and plaster, or fibreglass.

9.1.6. Filament Winding

This process consists of winding continuous filament over a suitably shaped mandrel. The filaments are in bundles which usually consist of thousands of individual fibres, and are referred to as rovings (fibreglass) or tow (carbon). The fibres are impregnated with resin just before they go onto the mandrel for the wet winding process (Fig. 9.7). There is

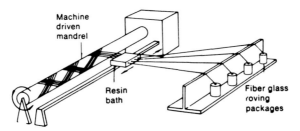

FIG. 9.7. Filament winding. (Courtesy of PPG Industries.)

also dry winding process; in this, prepreg tapes are used. Epoxy resins are often used for the matrix, though polyester and other resins may also be used.

The winding machines are of two types; rotating mandrel and stationary mandrel. Two winding methods are available: polar (or planar) winding, in which each layer of fibres is wound without spaces or cross-overs, and helical winding, in which both spaces and cross-overs occur. In both winding processes the fibres are laid onto the mandrel in a helical pattern, and the helix angle is chosen to suit the application. The arrangements of the fibres at the ends are most important for pressure vessels, and poorly designed fibre patterns can lead to early failure at the ends.

The construction of the mandrel requires considerable skill. It must not collapse under the large pressure resulting from the fibre-winding tension, and must be easily removed when the process is complete. Segmented metal forms (usually steel) are most commonly used, and they are faced with plaster.

This method produces very strong composites, and extremely large cylindrical and spherical vessels can be built.

9.1.7. Pultrusion

This is a method of making very strong aligned fibre composites. The fibres are impregnated with resin and pulled through a mould shaped to produce the desired cross-section in the product. The method is suitable for use with thermosets. The mould is heated to promote setting of the resin after it has impregnated the fibres. Further heat is applied either before the composite enters or after it leaves the mould. When the material has hardened it is cut into suitable lengths; see Fig. 9.8. Many types of sections are available from the pultrusion process.

9.1.8. Bag moulding

This method may also be used for making large parts. There are three different ways in which the moulding may be done, as shown in Fig. 9.9. The pressure bag system is relatively expensive since the combined mould-pressure vessel can only be used for one shape. The vacuum bag and the autoclave are very versatile, and relatively cheap.

Prepegs with thermosetting resins are normally used. Moulds are made from steel, aluminium, reinforced plastics, or plaster. Considerable skill on the part of the operator is required to achieve the full potentialities of the method.

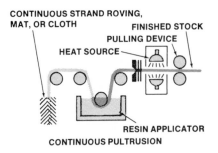

FIG. 9.8. Continuous pultrusion. (Courtesy of Fiberglas Canada.)

9.1.9. Matched Die Moulding

This method is used for the production of large number of identical parts. The moulds are expensive, but fast production rates can be obtained.

The reinforced plastic is cured between two heated mould surfaces, usually under high

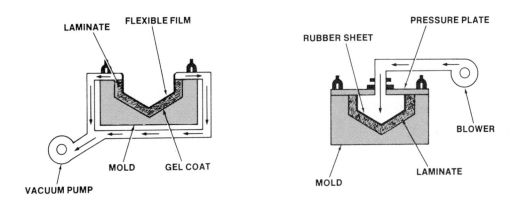

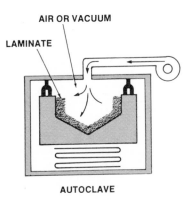

FIG. 9.9. Bag moulding methods: (a) vacuum, (b) pressure, (c) autoclave.
(Courtesy of Fiberglas Canada.)

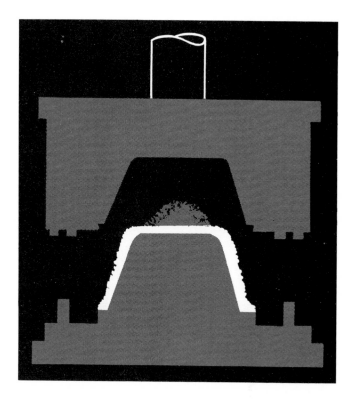

FIG. 9.10 Matched die moulding. (Courtesy of Fiberglas Canada.)

pressure, in a space of carefully controlled size and shape (Fig. 9.10). The fibres and matrix can be combined in a number of different ways prior to moulding. Four methods have already been described in Section 9.1.3: sheet moulding compounds, bulk moulding compounds, prepegs and pellets. In addition two others are often used:

(1) Wet fabric. The reinforcement, in fabric form, is impregnated with uncured resin immediately prior to moulding.
(2) Preform. The reinforcement is already moulded into approximately the correct shape, and the matched die moulding is used for distributing the resin and curing, without much change in shape.

Since considerable heat and pressure are used in this process, the moulds are usually made of metal. However, when moderate temperatures are used, the cheaper, flexible plunger moulds may be employed.

Cold-press moulding can be used with thermosets, the curing taking place inside the mould, or with reinforced thermophastic sheets which are heated just prior to insertion in the mould.

9.1.10. Laminates and Sandwich Construction

High performance composites are usually made by lamination, i.e. the fixing together of sheets of aligned fibre reinforced polymers (or materials). The polar method of filament winding is an example of this form of structure, but other methods of manufacture already described in this chapter can be used to produce the same structure form. A typical method of continuous laminate construction is shown in Fig. 9.11.

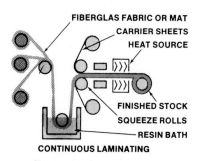

FIG. 9.11. Continuous laminating. (Courtesy of Fiberglas Canada.)

Normally each successive layer of the laminate has a different fibre direction from the previous one, except for the two layers at the centre. Usually the layers are "balanced", i.e. they consist of an even number of sheets, arranged so that the interface between the two sheets at the centre is a mirror plane of symmetry. This is to avoid unwanted twisting and other distortions which occur with unbalanced laminates when the laminate is stressed. The fibre directions used are chosen to suit the magnitudes and directions of the stresses that are expected to be encountered. Very high volume fractions of reinforcement can be obtained in laminates, and this is much the most efficient way of providing bi-directional or approximately transversely isotropic reinforcement. (Random fibres provide transversely isotropic reinforcement, but at the cost of low volume fraction and poor reinforcement efficiency.)

Sandwich construction is a method of obtaining the maximum stiffness in a structure when light weight is needed. This type of construction is widely used in aerospace applications, and examples are shown in Fig. 9.12. The outer skins may consist of high modulus, high strength composites, or aluminium, while the centre is a material of very low density, usually either a honeycomb type of structure, or a foam.

The sandwich is designed to be used in flexure. Thus the skin on one side must withstand tensile stresses, and the other skin must withstand compressive stresses. The filling of the sandwich needs only to resist shear stresses. Honeycomb fillings of reinforced plastic and aluminium provide the most efficient cores, having relatively large values of the ratio of shear strength to density, aluminium approaching 55 kNmkg^{-1}. Foams seldom exceed 10 kNmkg^{-1} (note that 1 MPa/(gm/cc) = 1 kNmkg^{-1} = 2.32 psi/(lb/cu ft);

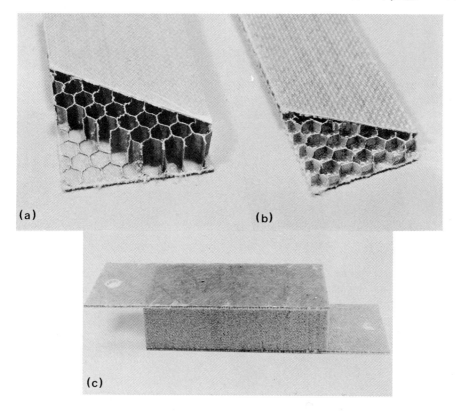

(a)

(b)

(c)

FIG. 9.12. Fibre-polymer sandwich constructions: (*a*) Kevlar-epoxy-aluminium honeycomb, (*b*)
Kevlar-epoxy-polymer honeycomb, (*c*) glass-polyester-polymer foam.

a solid piece of a strong aluminium alloy has a value of shear strength to density of about
$10 \, kNmkg^{-1}$).

A critical feature of both laminates and honeycomb structures is the adhesion between
the layers. Neither structure can operate efficiently without a very strong adhesive bond
between layers of components.

9.2. Properties

A very wide range of properties is possible, since there is a wide choice of possible
matrices and reinforcing materials. In addition, fibres may be chopped into various
lengths, and randomly oriented or organised in a wide variety of ways. Also, the matrix
may contain fillers and pigments, etc., which usually degrade its properties. (Fillers are
often used to reduce the price of the material, and it is interesting that chopped glass has
been used for this purpose, rather than as a reinforcement.)

In view of the wide range of possible properties this section will not contain an
exhaustive description of them, but rather will give a few significant values, and discuss
trends.

9.2.1. Strength and Modulus

Well-made composites obey the Rule of Mixtures for modulus (equation (4.12)) and the strength can slightly exceed the Rule of Mixtures for strength (equation (4.19)) due to the effect of flaws. Since the strength and modulus of most polymers are so low, compared with the fibres usually used for reinforcement, we can often ignore the matrix contribution and not be far wrong in assuming that well-made aligned fibre-polymers can have strengths of $V_f \sigma_{fu}$ and moduli of $V_f E_f$.

Commercial production does not meet these high standards, and we find that unidirectional glass fibre composites usually reach only about 65 % of the mixture rule for strength, although they do often reach the mixture rule for modulus.

With glass cloths, reinforcement is in two directions, with correspondingly less in each, so that only about 25 % of the law of mixtures value is obtained for strength, although the modulus reaches 65 % of the mixture law value. (Note that relative strengths in other than the fibre directions will be a little less than this.) With random mat the values are correspondingly less.

In addition, the highest volume fraction that can be obtained with random mat is only about 0.4, due to inefficiency of fibre packing. For square cloths V_f, can be 0.5, and with unidirectional fibres good composites can be made with $V_f = 0.8$. Table 9.2. summarizes these results.

TABLE 9.2. *Approximate Fraction of Rule of Mixtures Values Obtained for Strength and Modulus for Various Fibre Configurations, and Maximum Volume Fractions for Glass Reinforced Plastics*

Fibre configuration	Modulus[+] fraction	Strength[‡] fraction	Maximum V_f
Random mat	0.20	0.15	0.4
Cloth	0.65	0.25	0.5
Unidirectional	1.00	0.65	0.8

[+] Fraction of $V_f E_f$ achieved in commercial production.
[‡] Fraction of $V_f \sigma_{fu}$ achieved in commercial production.
(In both cases the matrix contributions are small enough to be neglected.)

Similar results are obtained with boron, carbon, and Kevlar, but often greater care is taken with the production of these much more expensive composites, so that the unidirectional strength fraction is usually close to 1.0 instead of 0.65. (Carefully made aligned glass reinforced plastics also have strength fractions of 1.0.)

The compressive strengths of glass fibre and carbon fibre reinforced polymers are usually close to, or somewhat less than, their tensile strengths. Boron reinforced polymers are generally about 20 % stronger in compression than in tension. This is probably due to the larger diameter of the boron. Kevlar fibre reinforced materials have poor compressive properties, owing to the weakness of the Kevlar fibres when compressed. For example, the compressive strength of $V_f = 0.6$ aligned Kevlar reinforced epoxy was only 0.28 GPa, while its tensile strength was 1.66 GPa. Kevlar reinforced polymers should thus not be used in situations where considerable compressive stresses have to be supported (for

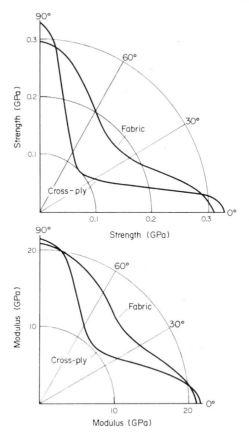

FIG. 9.13. Polar plots of modulus and strength of glass-epoxy laminates.

instance, in flexure, the $V_f = 0.6$ Kevlar reinforced epoxy only has a strength of 0.58 GPa).

The strengths and moduli of aligned fibre composites and laminates in directions other than the fibre directions are best represented graphically in the polar plots shown in Fig. 9.13. However, the stiffness and modulus matrices may also be used. Table 9.3 gives some data on aligned carbon-epoxy.

TABLE 9.3. *Elastic Constants and Strength Data for a Transversely Isotropic Aligned Carbon-Epoxy*

Compliances: (nPa^{-1})	$S_{11} = 0.0049$
	$S_{22} = S_{33} = 0.21$
	$S_{23} = -0.0060$
	$S_{12} = S_{13} = -0.0016$
	$S_{44} = S_{55} = 0.16$
	$S_{66} = 0.44$
Strengths: (GPa)	$\sigma_{1u} = 0.97$
	$\sigma_{2u} = \sigma_{3u} = 0.018$
	$\tau_{12u} = \tau_{23u} = \tau_{31u} = 0.055$

Large quantities of reinforced thermoplastics are manufactured with fibres chopped to very short lengths. These composites have very much poorer mechanical properties than those described above. However, they can be moulded with relative ease using injection and extrusion machines, so long as the fibre length is kept below 10 mm, and volume fractions of fibres are below 0.2. The composites produced have greater strength, higher modulus, and usually greater impact strength than the unreinforced matrix. In addition the thermal expansion is reduced, and properties are less dependent on temperature.

If cheap fibres (e.g. glass or asbestos) are used the short-fibre composite is often cheaper than the unreinforced matrix. There is usually little advantage in using the more expensive fibres (boron, carbon, and Kevlar) in these composites. Usually about a two-fold increase in strength and modulus can be obtained with glass or asbestos, and little better improvement can be obtained with the more expensive fibres. Both thermoplastics and thermosets are used in these moulding compounds.

9.2.2. Toughness

Although the Charpy and Izod impact methods of measuring toughness give only approximate values, most of the toughness measurements on composites have been made with these tests. This is because these tests are already well established, having been developed early in this century, and are simple to carry out.

With isotropic materials $G_c = K_{Ic}^2/E$, where E is the Young's modulus. Composites are highly anisotropic, and the corresponding relation for orthotropic composites is,

$$G_c = K_{Ic}^2 (S_{11} S_{22}/2)^{\frac{1}{2}} \left\{ \left[\frac{S_{22}}{S_{11}} \right]^{\frac{1}{2}} + \frac{2S_{12} + S_{66}}{2S_{11}} \right\}^{\frac{1}{2}} \tag{9.1}$$

for cracks propagating normal to the 1 axis, where the S's are the components of the compliance matrix (see equation (1.46)).

With non-fibre reinforced materials it is essential that the notch is very sharp (a sufficiently sharp notch is usually obtained with metals by machining a notch and then fatiguing the specimen). With composites this is not effective for notches propagated normal to the fibres, owing to the notch-blunting effect of the fibres.

Fracture mechanics tests have not so far been very successful with the tougher aligned fibre composites (e.g. glass reinforced polymers) due to difficulties in getting the crack to propagate in the desired direction. Representative toughness data taken from impact tests for aligned fibre composites are given in Table 9.4. It may be seen that the least tough fibre composites are the boron fibre reinforced ones, carbon being next, then glass, and the most tough being Kevlar. The relative brittleness of the carbon materials is probably due to the small diameter and high modulus of the fibres. The Kevlar composites are tough because of the toughness of the fibres.

The fracture toughness has been measured on a number of multi-directional laminates and cloth reinforced materials, Table 9.5. It can be seen that greater toughness is obtained with fibres aligned in the direction of the applied stress, and normal to the crack. Considerable reduction in toughness results from the propagation of cracks obliquely to the fibres, when the fibres are brittle, because the fibres bridging the cracks are bent as well as being under tensile stress (see Section 7.6). This effect is illustrated in Fig. 9.14 for carbon and silica reinforced epoxies.

TABLE 9.4. *Toughness Values for Reinforced Epoxy Resins Obtained from Impact or Work of Fracture Tests for continuous Aligned Fibre Composites with Fibre Volume Fractions of about 0.6*

Reinforcing fibre	Work of fracture (kJ m^{-2})
Boron	18 [†]
Carbon	40 [†] [‡]
S-glass	130 [†]
Kevlar	140 [‡]

[†] Work of fracture test.
[‡] Impact test.

TABLE 9.5. *Fracture Toughness (K_{Ic}) Values Taken from Notched Tensile Tests on Reinforced Epoxy Resins (All results given in MPa m$^{-\frac{1}{2}}$.).*

Reinforcing Fibre	Aligned fibres		Fabrics	
	0/90/ ±45	±45	0/90/ ±45	±45
Kevlar	29–40	11–18	22–32	17–26
E-Glass	17–23	10–17	15–20	12–17
Carbon	12	–	–	–

From Zweben, C. (1974) ASTM Symp. on Fracture Mechanics of Composites.

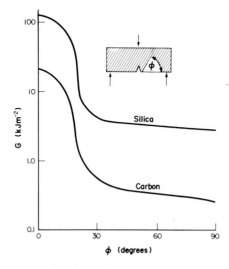

FIG. 9.14. Work of fracture as a function of fibre orientation for reinforced epoxy resins. (After Ellis, C. D., Ph.D. (1974) Thesis, Univ. of Bath.)

9.2.3. Fatigue

The fatigue properties of fibre reinforced polymers are good compared with aluminium. After 10^7 cycles, a typical aluminium alloy retains less than 20 % of its strength (see Fig. 1.4). Under the same conditions (equal tensile and compressive stresses during each stress cycle) carbon fibre reinforced epoxy, for example, retains 30 % of its strength, while under purely tensile fatigue conditions it retains more than 90 % of its strength. Boron reinforced polymers perform almost as well as this, while glass fibre reinforced polymers perform rather less well. Fig. 9.15 summarizes some results for fibre reinforced polymers. (Note that internal heat is produced when reinforced polymers undergo cyclic stressing. True fatigue effects, as distinct from effects due to heat, are thus only found at frequencies of less than 30 Hz.)

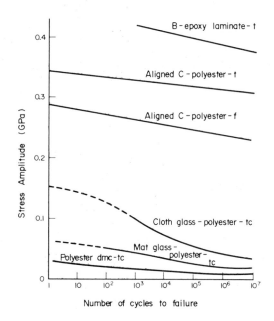

Fig. 9.15. An approximate comparison of fatigue behaviour. Modes of testing: t = tension, tc = tension–compression, and f = flexure. (After Harris, B. (1977) Composites **8**, 214). (dmc = dough moulding compound.)

The cause of the poorer fatigue resistance of glass reinforced polymers is the lower modulus of glass, and hence the greater strain suffered by the composite. Fibre–matrix bond failure is the first indication of fatigue; next the matrix starts to fail, and finally complete separation takes place. These processes are well illustrated by chopped strand mat glass fibre laminates, which has relatively poor fatigue properties (Fig. 9.16). Debonding has also been observed during shear fatigue.

The fatigue properties of composites are thus controlled by the maximum strain of the matrix. In addition, compression stresses during the fatigue cycle have a large effect on the fatigue life. This is illustrated in Fig. 9.17. A material which is unaffected by compressive stress would have the 10^6 cycles results in the form of two straight lines at angles of $\pm 45°$

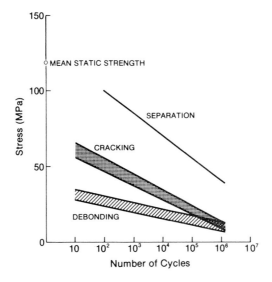

FIG. 9.16. Stages in the failure of chopped glass-polyester. (After Owen, M. J., Smith, T. R. and Dukes, R. (1969) *Plastics and Polymers* **37**, 227.)

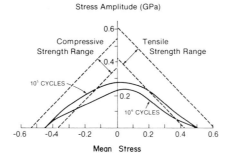

FIG. 9.17. Fatigue failure envelopes for carbon-epoxy laminates $0 \pm 45°$ orientation. (After Bevan, L. G. (1977) *Composites*, **8**, 227.)

to the abscissae, meeting at the ordinate at 520 MPa. The experimental results fall below the line by an amount depending on how symmetrical the stress cycle is.

The greater the proportion of fibres there are in the direction of the applied stress, the greater the fatigue life. This, again, is because the process is controlled by the matrix strain; the matrix strain will be kept to a minimum if the number of fibres directed along the stress axis is a maximum. The effect of the fibre volume fraction on fatigue life also depends on the matrix strain. Thus, for the same matrix strain, the fatigue life is independent of volume fraction. (The corresponding stress, of course, is related to the volume fraction by the Rule of Mixtures.)

Different lamination arrangements result in different matrix strains. Thus the results for one lamination arrangement cannot be used to predict the fatigue life of a different arrangement.

The fatigue properties are affected by notches. The presence of a hole reduces the fatigue life, and a slot reduces it even further, as shown in Fig. 9.18.

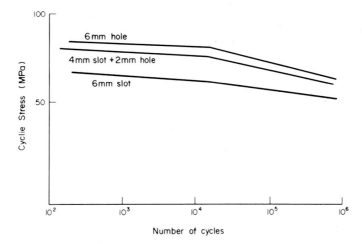

F IG. 9.18. Effect of different types of notch on glass-epoxy laminates. (After Carswell, W. S. (1977) *Composites* **8,** 251.)

9.2.4. *Creep Properties*

Polymers creep at room temperature at stresses far below their ultimate strengths. Figure 9.19 shows that at room temperature an epoxy resin creeps markedly at 34 MPa (and it still creeps significantly at 3.4 MPa). The addition of fibres with sufficiently great aspect ratios to give good stress transfer (i.e. s greater than $10s_c$) greatly reduces the creep of the polymer. If continuous ceramic fibres (boron, carbon, or glass) are used, and they are aligned in the stress direction (for uniaxial stressing) creep is almost eliminated

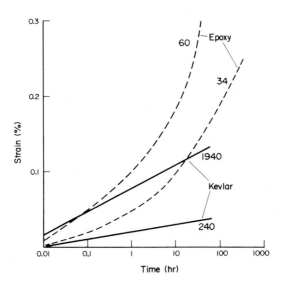

F IG. 9.19. Room temperature creep curves for Kevlar fibres and a cycloaliphatic epoxy resin. The applied stresses (MPa) are marked on the curves. (After Ericksen, R. H. (1976) *Composites* 7, 189.)

altogether. However, creep can still occur when the stress is not in the fibre direction, and can be quite substantial for relatively small angles between the stress axis and the fibre alignment axis.

When biaxial stresses are present, laminates with fibres in appropriate directions can be used to keep creep rates down to a very low level. There is some residual creep, however, because the polymer layers between the laminae can creep to a small extent.

In the case of glass fibre reinforced polymers, creep rupture occurs at stresses somewhat below the short-term ultimate tensile strength, even though no creep strain is observed. In the region near the breaking-stress, a 0.3 % increase in stress decreases the life by a factor of ten.

In contrast with ceramic fibres, Kevlar fibres do creep significantly, even at room temperature; Fig. 9.19 shows the creep of Kevlar fibres at two stresses at room temperature. Because of this, polymers reinforced with continuous Kevlar fibres creep at room temperature when stressed in the fibre direction. The creep curves for an epoxy resin containing 50 % Kevlar have regions of primary creep and secondary creep (in this region the creep strain rate is approximately constant). Results obtained, plotted as creep strain vs. log time, are shown in Fig. 9.20. These curves can be accounted for in terms of the creep of the matrix and fibres. The matrix controls the early stages of the creep process, while the fibres control the later stages. In addition, low stress creep depends on matrix creep, while creep at high stresses is governed by the fibre creep properties.

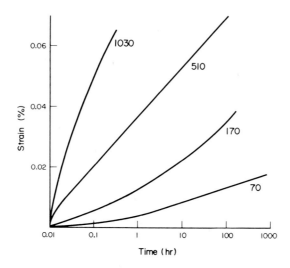

FIG. 9.20. Room temperature creep curves for Kevlar-epoxy. The applied stresses (MPa) are marked on the curves. (After Ericksen, R. H. (1976) *Composites* 7, 189.)

9.2.5. *Environment Effects*

Agents in the normal ambient environment which can degrade the properties of reinforced polymers include ultraviolet radiation, heat, and water.

Ultraviolet radiation and heat both degrade the polymer. Where a large intensity of ultraviolet radiation is present polymers are normally treated with materials which will absorb the radiation, which would otherwise cause chain scission. This solves the problem, but adds significantly to the cost of the polymer. The fibres in reinforced polymers usually absorb this radiation, so the problem is very much reduced, and the radiation only affects the very thin layer of polymer between the free surface and the fibre nearest the surface.

Heat also causes decomposition of polymers, and severely limits the usefulness of reinforced polymers above about 200 C. The maximum temperature depends on the polymer. The creep rate of short fibre reinforced polymers is greatly increased by heat. Figure 9.21 shows the temperature variation of the stress which, after being applied for 100s gives rise to a strain of 0.1 % for short glass fibre reinforced polypropylene. It is clear that the material loses a great deal of its stiffness when heated. Continuous fibre reinforced polymers are not so badly affected, but loss of properties occurs at the softening point of the polymer. Special polymers, most notably polyimides, have been developed to combat this problem. They extend the useful temperature range of reinforced polymers to about 300 C.

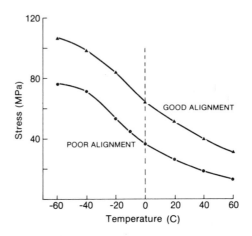

FIG. 9.21. Creep of chopped glass-polypropylene. Stress is that required to cause 0.1 % strain after 100s. (After Darlington, M. W., McGinley, P. L., and Smith, G. R. (May 1977) *Plastics and Rubber: Materials and Applications*, 51.)

At low temperatures polymers become brittle, but reinforced polymers continue to be able to support loads without loss of properties, except under conditions where multiple cracking of the polymer can occur. This has been observed in the case of steel wire-polymers, under stress, at liquid nitrogen temperatures, but glass-polymers have performed well at these temperatures.

The thermal stability of the composite is very much better than that of the unreinforced polymer. Thermal expansion coefficients for polymers and some composites are compared in Table 9.6. Some values for metals are also included, and it may be seen that the composites with high fibre loadings have coefficients which are considerably less than those of the less susceptible metals (e.g. cast iron).

Water affects both the polymer, and the polymer-fibre interface. In the case of Kevlar

TABLE 9.6. *Approximate Thermal Expansion Coefficients; μK^{-1} for Glass Reinforced Polymers (α_c) and the Unreinforced polymers (α_m) Compared with some Widely Used Metals, α_m.*

Reinforced thermosets	α_m	α_c	V_f
Polyester moulding compounds	50–100	14	0.15
Phenolic moulding compounds	25–60	8	0.12
Filament wound glass-epoxy	55–65	4	0.63
Pultruded glass-polyester	50–100	5	0.63
Reinforced thermoplastics	α_m	α_c	V_f
Polycarbonate, PVC[†] and Polyphenylene oxide	55–70	20	0.22
Polypropylene, ABS[‡], SAN[§] and Polystyrene	70–100	29	0.22
Polyethylene (medium density)	140–160	31	0.22
Acetal	81	34	0.22
Metals	α_m		
Cast iron and carbon steel	11		
Brass	21		
Aluminium	23		
Magnesium	27		
Zinc	28		

[†] Polyvinylchloride.
[‡] Acrylonititrile-butadiene–styrene copolymer.
[§] Styrene–acrylonitrile copolymer.

fibres it affects the fibres as well. Figure 9.22 shows the absorption of water at 100 C by carbon, glass, and Kevlar fibre reinforced epoxy. In the case of the glass and carbon, the absorption is about the same as that for the equivalent amount of resin, while Kevlar

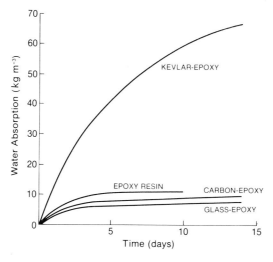

FIG. 9.22. Moisture absorption of epoxy resin and composites at 100 C. (After Phillips, D.C., Scott, J. M. and Buckley, N. (1978) Proc. ICCM2, 1544.)

absorbs much more. The absorbed water has a large effect on the shear strength of the Kevlar-epoxy, reducing it by 50%, but does not affect the shear-failure strain. With carbon- and glass-epoxy the shear strengths are not much affected, but the shear-failure strains are reduced by about 40%. Drying out the materials does not restore the properties. In addition the moisture affects the shear-fatigue properties of the glass- and carbon-epoxies, but not the Kevlar-epoxy. Figure 9.23 shows its effect on the fatigue curves for glass and carbon. The loss in fatigue strength is not recovered when the materials are dried out.

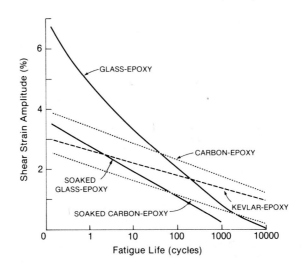

FIG. 9.23. Effect of water soaking for 7 days at 100 C on shear fatigue lives. The Kevlar-epoxy was not significantly affected. (After Phillips, D. C., Scott, J. M. and Buckley, N. (1978) Proc. ICCM2, 1544.)

9.2.6. Hybrid composites

The great advantage of composite materials for the designer is the versatility of the material. Its properties can be controlled by variation of the type, amount, orientation and length of fibre and type of matrix used, and to a limited extent by control of the interface. Thus, to a degree, the material can be designed to suit the application.

However, it was soon noticed that much better control of material properties could be obtained by making composites which included different types of fibre. Thus for optimum stiffness and toughness, a mixture of carbon and Kevlar is better than glass, or Kevlar alone. Figure 9.24 shows this effect. Laminates can be made in which each lamina is made with a different fibre; alternatively, the composite can be made by intimately mixing the fibres of different types. These are both hybrid composites. There have been suggestions that with mixed fibre composites, synergistic effects might be obtained. The toughness results in Fig. 9.24 seem to indicate this, since all the experimental points are above the lines joining the works of fracture of the respective non-hybrids. The evidence here is so far not conclusive, however. Little error will result from assuming that toughness of hybrids is given by the expression

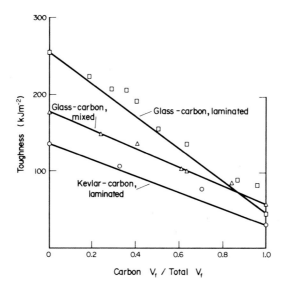

FIG. 9.24. Toughness of hybrids containing carbon. (After Dorey, G., Sidey, G. R. and Hutchings, J. (1978) *Composites* **9**, 25, Hancox, N. L. and Wells, H. (1973) ibid. **4**, 26, and Harris, B. and Bunsell, A. R., ibid. **6**, 197.)

$$\mathbf{G}_1 = (V_{fa}\mathbf{G}_{1a} + V_{fb}\mathbf{G}_{1b})/(V_{fa} + V_{fb}) \tag{9.2}$$

where $\mathbf{G}_{1a}$ and $\mathbf{G}_{1b}$ are the works of fracture of the non-hybrid composites and V_{fa} and V_{fb} are the volume fractions of the two fibre types. In this expression $V_{fa} + V_{fb} = $ constant. The expression simplifies if $\mathbf{G}_{1a}$ and $\mathbf{G}_{1b}$ are proportional to the respective volume fractions, given by $\mathbf{G}_{1a} = V_{fa}\mathbf{G}_{1a}^*$ and $\mathbf{G}_{1b} = V_{fb}\mathbf{G}_{1b}^*$. Then

$$\mathbf{G}_1 = V_{fa}\mathbf{G}_{1a}^* + V_{fb}\mathbf{G}_{1b}^*. \tag{9.3}$$

The Young's modulus of the hybrid is also given by an expression like equation (9.3). Thus

$$E_1 = V_{fa}E_{fa} + V_{fb}E_{fb}. \tag{9.4}$$

Hybrid composites obey this expression well. If the total volume fraction of fibres, $V_{fa} + V_{fb}$, is small, we should add $V_m E_m$ to the right-hand side of equation (9.4).

The compression strength of hybrids has not been investigated in detail, but with carbon–Kevlar laminated hybrids does appear to be given by a Rule of Mixtures expression (Fig. 9.25).

The flexural properties depend on how the composite is made. Rule of Mixtures expressions do give a reasonable representation for some intermixed hybrids, and even some laminated hybrids (Fig. 9.25). Other laminates do not obey the Rule of Mixtures for strength or modulus in flexure. One reason for this is that the stressing system is complex in a hybrid laminate. If the laminate is to be flexed it is advantageous to put the stiffer fibre at the surfaces to promote maximum stiffness. However, this fibre is then subject to a higher stress, and early failure can be expected. The design of laminates for use in flexure must be governed by the failure strains of the component fibres.

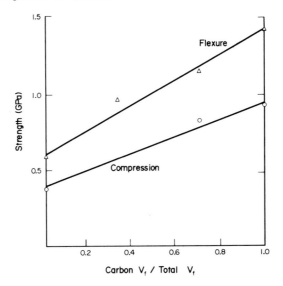

FIG. 9.25. Carbon–Kevlar–epoxy laminated hybrids. (After, Dorey, G., Sidey, G. R. and Hutchings, J. (1978) *Composites* **9**, 25.)

The tensile strength of a laminate is also controlled by the failure strains of the fibres. Figure 9.26 shows that tensile strengths generally lie below the Rule of Mixtures. We expect the lower breaking-strain fibres to break first. This does not necessarily precipitate

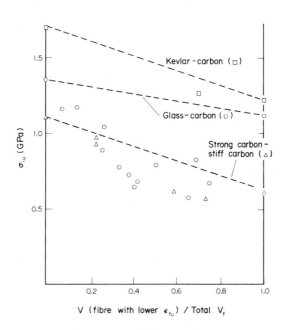

FIG. 9.26. Tensile strength of hybrids. (Data from Aveston, J. and Kelly, A. (1979) *Phil. Trans. Roy. Soc.*, Dorey, G., Sidey, G. R., and Hutchings, J. (1978) *Composites* **9**, 25, and Edwards, H. E., Parratt, N. J. and Potter, K. D. (1978) Proc. ICCM2, 975.)

composite failure. These fibres can, in some circumstances, break up into their critical lengths, when they can still contribute reinforcement to the extent of $0.5\ V_{fa}\sigma_{fua}$. The composite strength is thus

$$\sigma_{1u} = 0.5\ V_{fa}\sigma_{fua} + V_{fb}\sigma_{fub} \qquad (9.5)$$

when V_{fb} is relatively large; see Fig. 9.27. When $V_{fb} = 0$, the composite strength is $V_{fa}\sigma_{fua}$. The composite fails when $\varepsilon_1 = \varepsilon_{fua}$. Thus when there are small amounts of the higher breaking-strain fibre we have an expression of the type

$$\sigma_{1u} = V_{fa}\sigma_{fua} + V_{fb}E_{fb}\varepsilon_{fua} \qquad (9.6)$$

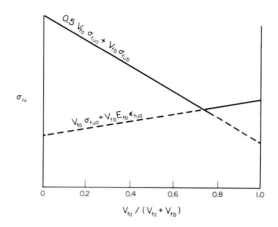

FIG. 9.27. Expected variation of strength for a hybrid of fibres with different breaking strains. V_{fa} is the volume fraction of the lower breaking-strain fibre. $V_{fa} + V_{fb} = $ constant.

This is also shown in Fig. 9.27. (This graph was plotted for the combination of stiff carbon and strong carbon, Table 3.2.) Ideally the strength will be given by equation (9.5) or (9.6), whichever is the greater. However, in practice there is no guarantee that the composite can withstand the breaking up of the lower breaking-strain fibre. Thus equation (9.5) is an upper bound and equation (9.6) is a lower bound to the tensile strength of hybrids. Unfortunately we seldom know the values of σ_{fua} and σ_{fub} in the composite with any accuracy. There are two reasons for this. One is the damage that the fibres suffer during composite manufacture. The other is that the strength of the fibres having the critical length is seldom known. This is generally greater than the strength as normally measured (see Section 6.11).

9.3. Platelet Reinforcement

Platelets have the advantage of endowing stiffness in two directions, when they are suitably aligned in a polymer. Mica, aluminium diboride, and glass reinforcements have been intensively investigated. It has been found that parts can be made using a wide variety of moulding techniques, and near the surfaces of the parts the platelets are usually aligned parallel to the surface. However, near the centre the alignment is often normal to the surface (especially with thick sections) so the tensile strength is sometimes less than half the flexural strength.

Before being used to reinforce a polymer it is essential to process mica to separate out the low aspect ratio pieces. Otherwise strengthening may be negligible. Mica is not very hard, and so does not abrade the processing equipment as much as fibreglass does.

The mechanical properties of platelet reinforced polymers are not as good as their fibre counterparts. Some data for reinforced epoxies are given in Table 9.7 for $V_p = 0.6$. Aluminium diboride is the most promising from this point of view, but lack of toughness is a big drawback. Properties increase linearly with V_p up to about 0.6. Above 0.6 they decline.

TABLE 9.7. *Average Properties of Platelet Epoxies at* $V_p = 0.6$

Platelet	Strength (GPa)			E	Toughness	Density
	Tensile	Flexural	Compressive	(GPa)	$(J\,m^{-2})$	$(Mg\,m^{-3})$
Mica	0.076	0.20	0.13[†]	48	34	2.0
Al B$_2$	0.45	0.64	0.45	293	90–4000	2.27
Glass	0.16	0.29	0.34	41	–	2.05

[†]Mica-thermoset.

Mica is finding some uses in non-load-bearing situations, since it imparts stiffness at very low cost (raw mica costs about 0.1\$ kg^{-1}, and after processing to obtain aspect ratios of 80–100 costs about 0.25\$ kg^{-1}). In thermoplastics it has to be processed with great care.

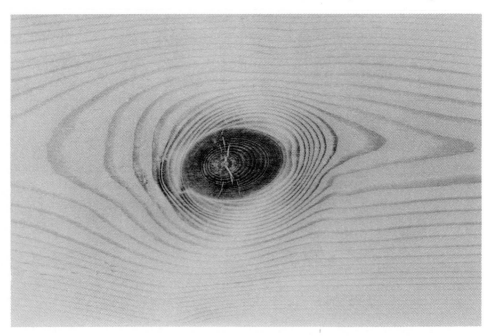

FIG. 9.28. Nature's joint. Note how the grain changes around the joint, indicating a change in fibre orientation.

The degradation can be seen from the decrease in flexural modulus of mica-polystyrene, comparing compression moulding, injection moulding, and extrusion followed by injection moulding. The respective moduli are 41, 37, and 25 GPa. The corresponding strengths are 0.12, 0.14, and 0.07 GPa.

Hybrid composites have been made using chopped glass fibres to enhance the toughness of the mica-polymer. Thus, for instance, the toughness of reinforced thermoplastics increases by 50% if 11% of mica (by volume), and 16% glass is used instead of just 27% mica. The glass also improves the tensile strength by about 20%, but the modulus decreases by about 10%.

9.4. Joints

In a complicated structure it is usually necessary to divide the structure into more elementary parts, construct these, then join them together.

Composites present special problems for joining because of their extreme anisotropy. Even cross-ply laminates have problems because of their low out-of-plane properties.

Nature has solved the problem (Fig. 9.28) by so arranging fibre growth around the joint that it has optimum properties. This approach is not usually available to designers of composites.

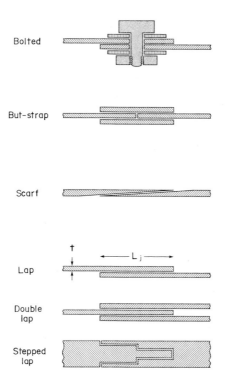

FIG. 9.29. Types of joint suitable for joining laminates and sheets.

The most cost-effective methods of joining isotropic materials are usually either bolts or rivets. Rivets are seldom suitable for composites, but bolts may sometimes be cost effective in joining laminates to each other, or to sheets of other materials. Due to the low shear strength of the composite the bolt clamping load should be spread over a large area by the

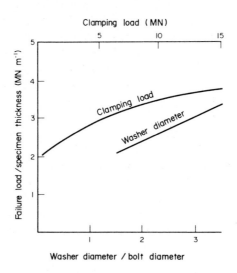

FIG. 9.30. Effect of clamping load and washer diameter on the load-carrying capacity of a bolt in a glass fibre laminate. (After Stockdale, J. H. and Matthews, F. L. (1976) *Composites* **7**, 34.)

use of washers, as shown in Fig. 9.29. The washer should be as large as possible, consistent with weight requirements. Figure 9.30 shows that the strength of the joint is a linear function of washer diameter. It is usually an advantage to have the clamping load high, but the maximum is limited by the material. Figure 9.30 shows that with glass-epoxy the joint strength is greatly increased by good clamping, though there is little to be gained by having the load greater than about 15 MN.

Bolted joints should be well fitting for maximum effectiveness. They should not be too close to the edge of the sheet, but about five bolt diameters (Fig. 9.31) is usually sufficient. They should only be used with extreme caution for unidirectional fibre composites, though, since they can very easily shear out under load.

Glued joints are normally used for unidirectional fibre composites. Butt-strap, scarf, and lap joints of various types are the main forms used for unidirectional composites and laminates. Examples are shown in Fig. 9.29.

In all these joints the length of the joint is calculated assuming that the adhesive has a shear strength τ_{au}. Thus, for sheets of thickness t, and a joint length of L_j, the single lap joint can transmit a stress σ_{ju} where

$$\sigma_{ju} = L_j \tau_{au}/t. \tag{9.7}$$

Equation (9.7) only works, however, for thin laminates, as shown in Fig. 9.32. Stress concentrations occur in this type of joint, which can be reduced by chamfering the edges.

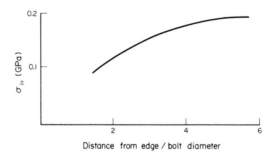

FIG. 9.31. Composite stress at joint failure (σ_{ju}) vs. distance of bolt from edge for boron-epoxy laminate/aluminium joint. (After Grimes, G. C. and Greimann, L. F. (1974) (Ed. Chamis, C. C., Academic Press) (*Composite Materials* **8**, 135.)

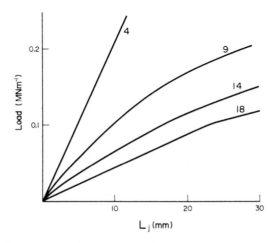

FIG. 9.32. Load on each ply per unit width of joint as a function of lap length, L_j, for glued double-lapped glass-polymer/aluminium joints. The figures on the curves indicate the number of layers in the laminates. (After Grimes, G. C. and Greimann, L. F. (1974) (Ed. Chamis, C. C., Academic Press) *Composite Materials* **8**, 135.)

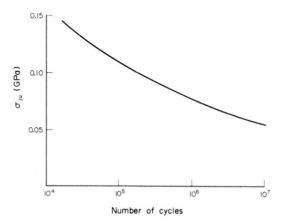

FIG. 9.33. Fatigue of unidirectional carbon-epoxy/aluminium glued $2\frac{1}{2}°$ scarf joint. (After Smith, M. A. and Hardy, R. (1977) *Composites* **8**, 255.)

However, these become serious when laminates with more than four layers are joined in this way. Thus the effective value of τ_{au} is reduced to a small fraction of its value for thin laminates. In this case the butt-strap, scarf, and single or double lapped joints should not be used. Instead the stepped lap should be employed.

Glued joints are affected by cyclic stresses, and show the type of fatigue behaviour usually observed with composites. Figure 9.33 shows that a joint strength decreases monotonically with number of stress cycles, without any indication of a fatigue limit. This was a scarf joint between unidirectional carbon-epoxy and aluminium.

When a laminate has to be fixed to a massive support a different type of joint is required.

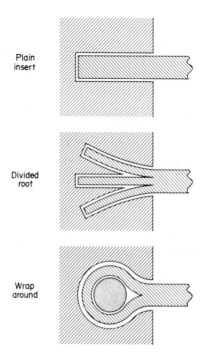

Plain
insert

Divided
root

Wrap
around

FIG. 9.34. Joints to connect laminates with massive support.

Some examples are shown in Fig. 9.34. The plain insert is the simplest and weakest. The divided root is much better, and the strength is greatly influenced by the number of divisions. The wrap around is considered to be very promising, but is difficult to construct.

Further Reading

BILLMEYER, F. W. (1971) *Textbook of Polymer Science* (Wiley-Interscience, New York).
LUBIN, G. (ed.) (1969) *Handbook of Fiberglass and Advanced Plastics Composites*, (Van Nostrand Reinhold, New York), Chapters 2–5, and 13–20.
RICHARDSON, M. O. W. (ed.) (1976) *Polymer Engineering Composites* (Applied Science Publishers, Elsevier).

10

Reinforced Metals

THE GREAT interest in fibre reinforcement that blossomed in the early 1960s was largely due to the feeling that conventional metallurgy was reaching its limit, and the new materials needed for the space age required a radical approach. The extraordinarily high strength and modulus of some ceramic whiskers, coupled with their excellent temperature and corrosion resistance, seemed to provide the answer. All that was required was a suitable medium to hold the whiskers in place at the high temperatures required by the new machines being designed. Metals seemed the obvious choice for this.

Unfortunately, despite a tremendous effort on the part of hundreds of scientists and engineers, the goal of a strong, light, tough, stiff, corrosion-resistant and thermally stable materials for use at higher temperatures than presently available metals (greater than 1100C) seems almost as far away as ever. Whisker reinforcement of metals is difficult and expensive, and the resulting composites are usually severely degraded by but a few thermal cycles.

Two different approaches seem more likely to be fruitful for the production of high-temperature materials in the immediate future: the use of refractory wires (e.g. tungsten), and the *in situ* composites, i.e. composites in which fibres or platelets are grown by controlled crystallization from eutectic and off-eutectic alloys.

The interest in reinforced metals has shifted its focus somewhat. It is appreciated that worthwhile improvements in high-temperature properties can be obtained even with some weight penalty; in addition, useful reinforced metals can be produced using continuous fibres of boron or carbon, for lower temperature applications.

Reinforced metals have a number of potential advantages over reinforced polymers, apart from their generally higher temperature of operation:

1. The ductility and higher strength of the metal matrix may give better off-axis properties and interlaminar shear strengths.

2. Joints are a matrix-controlled property; the greater strength and ductility of the metal, and the possibility of diffusion bonding of the matrix, make stronger joints possible than with reinforced polymers.

3. The modulus of the matrix is much higher than that of a polymer, so that at stresses below the matrix yield point, higher moduli are obtained than with reinforced polymers.

4. The erosion resistance, and water and organic solvent resistance of reinforced metals is greater than that of reinforced polymers. In addition metals are less permeable to gases and liquids.

5. The much greater electrical and thermal conductivity of the metal matrix can be an advantage in some applications.

There are some disadvantages, however:

1. The stress–strain curve has a "knee" at the matrix yield strain, so that the material is

truly elastic only over a small range of stresses, unless a hard matrix is used. Above the knee the effective elastic modulus is not much greater than for an equivalent reinforced polymer.

2. Reinforced polymers withstand certain corrosive environments much better than reinforced metals.

3. Reinforced metals are difficult to make, and hence usually much more expensive than reinforced polymers.

4. Reinforced metals have higher densities than reinforced polymers.

This chapter will be mainly concerned with metals having improved load-bearing properties resulting from the addition of fibres (or whiskers). The description of manufacturing methods includes a brief discussion of the important properties of metals affecting manufacture. This will be followed by a description of the properties of reinforced metals.

10.1. Methods of Manufacture

10.1.1. The Matrix

Metals are crystalline solids, and their mechanical properties depend a great deal on the perfection of the crystals. Large, nearly perfect crystals of pure metals are very soft and ductile at a small fraction of their absolute melting-temperature, T_m (for example pure iron is easily deformed at room temperature). They are hardened by cold working or by the presence of impurities, either as dispersed atoms or as crystals. The science of physical metallurgy is based on the control of crystal size and perfection, and the use of precipitates, in the making of alloys. An enormous range of properties can be achieved by this means. Some precipitates in traditional alloys behave in a way that is similar to fibre reinforcement; for example, Fe_3C precipitates in steel consist of fine platelets.

After a metal matrix composite has been put together the matrix can be extensively modified by cold work, or by heat treatment to cause precipitation of components of the metal alloy. Care has to be taken to prevent damage to the fibres, however.

Most metals are very reactive. In air they react readily with oxygen at 20 C, but some quickly form a very thin impermeable oxide film that inhibits further reaction. Good examples of this are Al, Cr, Mg, and Ni. When the oxide is permeable, reaction is slow at 20 C, unless suitable conducting paths are provided so that electrochemical processes are promoted. This happens in steel, where the presence of water and chloride ions promote rusting. At higher temperatures reaction rates are higher, but the impermeable oxide films usually maintain their effectiveness. Thus, for example, aluminium can be melted in air without much oxidation occurring.

The reactivity of metals is a serious problem in metal matrix composites, and the fibres often have to be protected by barrier layers, especially if the material is required for high-temperature operation.

Other aspects of the reactivity of metals can be put to good use. For example, metals can be electrodeposited from solutions of their ions, or deposited by chemical vapour deposition from gaseous organic compounds.

Metals are also fusible, though the melting-points range from -39 C (mercury) to 3410 C (tungsten). Melting-points can be reduced by alloying, and liquid infusion from the melt is an important method of composite manufacture. However, metal alloys are quite soft and easily worked at about 0.8 T_m. Thus, instead of melting it, the alloy can be moulded around the fibres by pressing at the appropriate temperature.

10.1.2. General Considerations

The choice of method used for making a composite is limited by the need to consider the following factors.

1. The adhesion between fibres and matrix must be adequate, so that the composite achieves the good shear and off-axis properties normally required of reinforced metals. Large numbers of wetting studies have been carried out to investigate the adhesion phenomenon. A widely used criterion is that the contact angle between a molten drop of the matrix material and the fibre material must be less than 90°. When the contact angle is greater than 90° the matrix may be modified by alloying, or the fibre may be coated with an interfacial layer that is compatible with both fibre and matrix (i.e., has a small contact angle with both fibre and matrix materials).

2. The interaction between fibres and matrix needs to be limited, so that the fibres are not weakened. The reaction can result in complete decomposition of the fibre material (this happens with silica reinforced aluminium) or in the formation of brittle layers of reaction product at the fibre surface (this happens with tungsten reinforced nickel). Partial or even slight decomposition of the fibre is usually disastrous, since the fibre strength is usually lost after relatively little surface attack. Brittle layers of more than about one-hundredth of the fibre diameter may have a very serious weakening effect, since they crack on cooling the composite after manufacture (or due to thermal cycling thereafter) and the cracks can easily propagate into the fibres when the composite is stressed.

3. The need to minimize mechanical damage to the fibres. In hot-pressing too much heat leads to too great an interaction, while too much pressure can cause severe mechanical damage to the fibres. The temperatures and pressures used are a compromise between excessive damage of one sort or the other, and inadequate densification.

4. The volume fraction of voids must be kept as small as possible. Voids are usually concentrated at the fibre–matrix interface, and weaken the composite by acting as stress-raisers and sites for initiation of cracking and debonding.

5. The achievement of the desired fibre geometry. It is seldom worth while to reinforce metals with randomly oriented fibres, since the small strengthening achievable (due to the relatively low volume fractions possible) is often not worth the trouble and expense. Thus, the manufacturing methods available are usually limited to those which produced aligned fibres.

Manufacturing methods can be classified according to the initial state of fibres and matrix. The fibres do not need to exist initially; they can be produced by extrusion from pellet-shaped particles within the matrix, or they can be generated by controlled precipitation from an alloy melt (this can produce either platelets or fibres depending on the volume fraction of the nascent reinforcement). The matrix in both cases is in the form of a billet or bar. When pre-existing fibres are used, the matrix can be solid, as sheets or powder, or liquid, or atomically divided, as vapour or ions. In the account that follows, these methods will be classified according to the initial state of the matrix.

10.1.3. Bar Matrix

We will first consider the co-extrusion of two phases. In this case, the matrix is the continuous phase, and the fibres result from the extrusion of a discontinuous phase of pellet-shaped particles. Both phases have to be sufficiently ductile so that they can be extruded without cracking. The method is therefore normally used to reinforce one metal with another metal. However, some oxides are ductile at high temperatures, and can be co-extruded with a metal, when hot, to produce oxide fibres in refractory metals. In this way, niobium and tantalum can be reinforced with magnesia, thoria, or zirconia. (These composites have roughly the same strength at high temperatures as the unreinforced metals, but are lighter and have higher moduli.) The reduction in cross-section during the extrusion process has to be quite large in order to produce fibres with sufficiently high aspect ratios to be efficient reinforcers.

The other method of making composites from matrix bars containing the fibre generating material is the controlled solidification and crystallization process. These are called *in situ* composites since the fibres are formed *in situ* during the solidification process. The bar can either be an off-eutectic or eutectic alloy. The discontinuous phase can either be rod-shaped or plate-shaped, rods usually being produced at volume fractions of 30% or less, and platelets at higher volume fractions.

The binary (and pseudo binary) eutectics are the most straightforward of this class of material. Figure 10.1 shows the phase diagram for the nickel–chromium binary alloys. An alloy containing 50 atomic per cent Ni and 50 atomic per cent Cr, when cooled from 1500 C say, will start to precipitate crystals which consist mainly of nickel (the γ phase) at about 1360 C, and continue to do so until the temperature reaches 1345 C. At this point, the remaining liquid mixture will solidify around the γ crystals produced during cooling between 1360 C and 1345 C. A solution of 40% Ni and 60% Cr on the other hand, when cooled from 1500 C will start to precipitate crystals which consist mainly of chromium (the α phase) at about 1400 C, and continue to do so until the temperature has fallen to 1345 C, when the remaining liquid mixture will solidify. However, a mixture of 45% Ni and 55% Cr will remain completely liquid as it is cooled, until the temperature has fallen to 1345 C, at which point it will solidify completely as a mixture of crystals of Cr principally (α) and nickel principally (γ). This is the eutectic mixture.

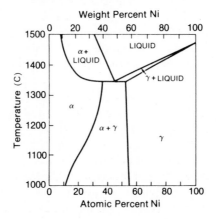

FIG. 10.1. Phase diagram for nickel–chromium.

Only the γ phase is produced when complete solidification takes place for alloys containing more than 53% nickel.

Directional solidification can be achieved with a bar of the eutectic mixture by putting it in a crucible which passes slowly through a heater into a cool area. The bar is melted, and as it passes into the cool area it freezes again, as indicated in Fig. 10.2. During the solidification the chromium crystallizes as rods in a matrix of nickel.

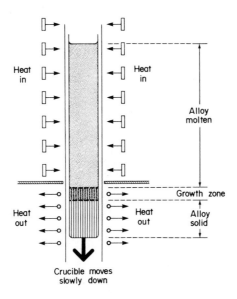

FIG. 10.2. Schematic view of crystal growth apparatus.

Not all eutectics form platelet or rod-like precipitates; the phenomenon is governed to some extent by the entropy of fusion of the components. Most metals and metal alloys do so, however.

With binary (and pseudobinary) eutectics, the volume fraction of the discontinuous phase is fixed, once the components of the alloy are decided upon. There are, however, many different alloy systems which can be used if pseudobinary, and certain ternary eutectics are included; for example, over 150 well-characterized alloys are available, based on nickel or cobalt. Most of these are potentially useful at high temperature, forming rod or platelet precipitates, and having volume fractions ranging from less than 10% to more than 60%. Figure 10.3 shows an *in situ* composite with fibres for the reinforcing phase.

Variations in volume fraction can be obtained by using off-eutectic alloys. These require much more stringent growth conditions, however, which is a great disadvantage. In addition, the matrix phase has a fixed composition, so that modifications to the matrix to improve its properties (for example, corrosion resistance) are not possible.

Variation of volume fraction can be obtained with ternary and quarternary alloys which have eutectics with a wide range of compositions. The crystallization conditions do not have to be so closely controlled to get the desired results with these alloys, but the phase diagrams are very complex.

The melting is usually carried out with induction furnaces, or furnaces with large heat

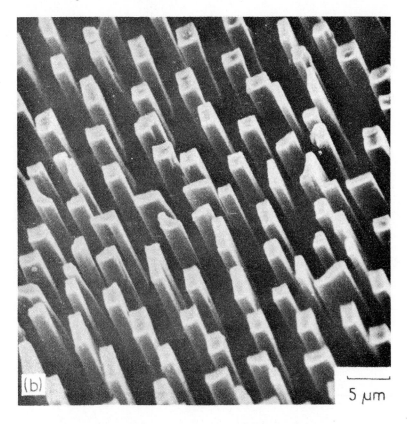

FIG. 10.3. Molybdenum fibres grown *in situ* in a Ni$_3$Al matrix, (Sprenger, H., Richter, H. and Nickl, J. J. (1976) *J. Mater. Sci.* **11,** 2075; courtesy Chapman and Hall.)

capacity. The alloy bar is usually held in a vertical crucible, melted, and then withdrawn from the furnace at a controlled rate, into a water-cooled region to ensure a sufficiently fast cooling rate, as shown in Fig. 10.4.

The zone-melting technique is also widely used. A molten zone is produced by induction heating, and the bar is slowly pulled through the induction heated so that the molten zone moves along the bar.

Fine structures (i.e. thin platelets or rods) are produced with fast cooling rates in the molten zone, while coarse structures are obtained with slow cooling rates. Great care is needed to keep the conditions constant; small variations can upset the crystallization process and result in the sudden termination of all the rods and platelets. (They restart further down the bar.) The region lacking the rods or platelets are usually very weak.

10.1.4. Sheet Matrix

In this method layers of fibres are laid between thin sheets of matrix foil in a mould, and the material is then consolidated by hot pressing. The process is shown schematically in

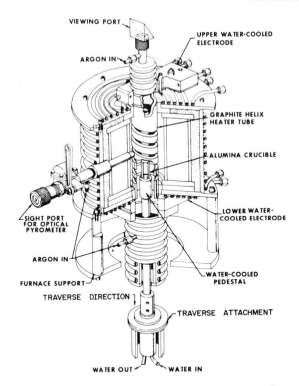

FIG. 10.4. Furnace used for growing *in situ* composites. (Thompson, E. R. and Lemky, F. D. (1974) *Composite Materials* **4**, 122, Kreider, K. G. (Ed.) Courtesy of Academic Press.)

Fig. 10.5. The method is only suitable for relatively large diameter fibres (e.g. boron) or wires, or for composite rods (made, for example, by pre-impregnating carbon fibre tow with matrix materials by one of the methods to be described later).

The temperature and pressure have to be controlled very carefully to ensure adequate consolidation without too much chemical interaction, or mechanical damage.

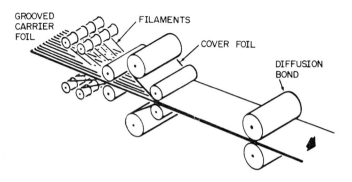

FIG. 10.5. Sheet matrix method for making fibre reinforced metals. (Metcalf, A. G. (1974) *Composite Materials* **4**, 293, Kreider, K. G. (Ed.) Courtesy of Academic Press.)

Tapes can be made up with sheet matrix on either side of a layer of fibres, the layers being held together by a resin binder (e.g. polystyrene) that evaporates during the first stage of the consolidation process. The tape can be used in the same way as polymer prepeg tape, in such processes as filament winding, etc.

10.1.5. Powder Matrix

In this method the fibres and matrix powder are combined, and held together by a volatile solid binder. When the mixture is hot pressed, the binder escapes by evaporation.

**HOT ISOSTATIC PRESS
SCHEMATIC DRAWING**

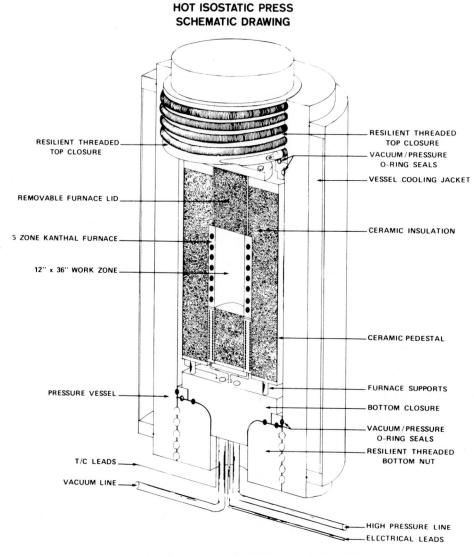

RESILIENT THREADED TOP CLOSURE

REMOVABLE FURNACE LID

5 ZONE KANTHAL FURNACE

12" x 36" WORK ZONE

PRESSURE VESSEL

T/C LEADS

VACUUM LINE

RESILIENT THREADED TOP CLOSURE

VACUUM / PRESSURE O-RING SEALS

VESSEL COOLING JACKET

CERAMIC INSULATION

CERAMIC PEDESTAL

FURNACE SUPPORTS

BOTTOM CLOSURE

VACUUM / PRESSURE O-RING SEALS

RESILIENT THREADED BOTTOM NUT

HIGH PRESSURE LINE

ELECTRICAL LEADS

FIG. 10.6. Schematic drawing of hot isostatic press used to make W–Ni alloys. (Courtesy of Westinghouse Canada Ltd.)

Often the pressing takes place in two stages. Stage one results in removal of the binder, and sufficient consolidation to hold the matrix and fibres together. Stage two involves pressing techniques similar to those described for reinforced polymers. The temperatures and pressures used are generally much higher for metals. This process consolidates the material to the maximum practical extent consistent with acceptable chemical and mechanical damage of the fibres. Figure 10.6 is a schematic drawing of the apparatus used. Reinforced polymer makers use different terminology from metallurgists here. Complex shapes are made by autoclave bag moulding reinforced thermosetting resins; the same process, when used by metallurgists, is called hot isostatic pressing.

10.1.6. Liquid Matrix

In this section the use of a molten matrix to surround and embed pre-existing fibres and whiskers will be described (the manufacture of *in situ* composites, which also involve a matrix liquefaction stage is described in Section 10.1.3.)

The simplest method of liquid infiltration is to pour the molten matrix into a vessel containing the fibres or whiskers. Whiskers can be in the form of a felt or mat, while fibres can be aligned in a tubular mould, fitting loosely in it. The method is not suitable when the matrix reacts strongly with the fibres (e.g. aluminium and silica). In addition, difficulties may be experienced with the wetting of the fibres or whiskers, though these can often be overcome by coating them. For example, sapphire whiskers have been successfully infiltrated with silver and aluminium by coating them first (by sputtering or condensation from the vapour) with more refractory metals such as nickel or nichrome. The method is less successful when used for infiltration with nickel.

In the case of mutually reactive matrix–fibre coupling (for example, aluminium and silica) each individual fibre may be coated by drawing it singly through a bead of the molten metal. Continuous single fibres can be moved very quickly through small metal beads to give adequate coating thicknesses with very little time for the chemical reaction, since the coating on a single fibre will cool extremely quickly. With this method, the coated fibres have to be hot pressed to make the composite.

With less reactive systems (for example, carbon–aluminium) a fibre tow, which can contain as many as 10,000 individual fibres, can be drawn through a crucible containing the molten metal. The fibres enter at the top, and pass through a die at the bottom of the crucible. The fibres can be protected against attack to some degree by a suitable metallic coating (nickel was used for carbon–aluminium) or the rate of reaction can be reduced by alloying the metal to reduce its melting-point so that the process can be carried out at lower temperature (for example, 12 % silicon in the molten aluminium matrix reduces the temperature of the operation from just over 660 C to 580 C). When the reinforced rods produced by this process are not thick enough for immediate use in a structure, a large number can be combined by hot pressing to produce the cross-section desired. Figure 10.7 shows three methods of infiltration to form a solid rod, and Fig. 10.8 shows carbon tows infiltrated with aluminium.

Another method that uses the matrix in liquid form during part of the manufacturing process is plasma spraying. The matrix is sprayed onto the fibres, which are supported on a foil, also made of the matrix material. The resulting tape is very porous, easily deformable, and suitable for cutting to the shape and size required for hot-pressing to the

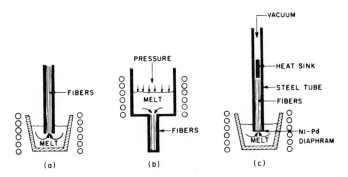

Fig. 10.7. Methods used for liquid metal infiltration. (Mehan, R. L. and Noone, M. N. (1974) *Composite Materials* **4**, 194, Kreider, K. G. (Ed.) Courtesy of Academic Press.)

desired form. One advantage of the plasma-spraying technique is that the metal rapidly cools and freezes in contact with the fibres, thus minimizing undesirable chemical reactions.

10.1.7. *Atomic and Ionic Matrix*

The matrix can be deposited onto the fibres directly from the vapour, or indirectly by a vapour phase reaction. (CVD: this process is used to make boron and SiC fibres; see Section 3.3.3). Aluminium has been coated successfully on carbon fibres by thermal decomposition of triisobutyl aluminium, while nickel carbonyl is used for the deposition of nickel.

Electroplating can also be carried out on any conducting fibre or wire. Air and water must be excluded for the deposition of aluminium, so ether is used as solvent. Aqueous baths are used for the more noble metals (copper or nickel for example).

The most serious difficulty with the techniques involving the matrix in these forms is that of ensuring good penetration of the matrix into fibre bundles. High rates of production are required for economic reasons, so that with carbon fibres, for example, it is necessary to plate the fibres in the form of tow rather than singly. However, deposition on fibres at the centre of the tow is prevented because of shielding by the outer fibres. Consequently, the tow has to be rearranged so that all the fibres are in a plane, and none are touching.

This problem is less serious with chemical vapour deposition and electroplating.

The coated fibres are hot pressed to produce the final composite. An organic binder may be used to make the coated fibres into prepreg tapes to ease subsequent production steps.

10.1.8. *Laminates*

Fibre reinforced metals can be made by lamination in the same way as for fibre reinforced polymers. The simplest way of doing this is to use the various tapes assembled in the desired orientation and hot press them. The pressure, temperature, and time of

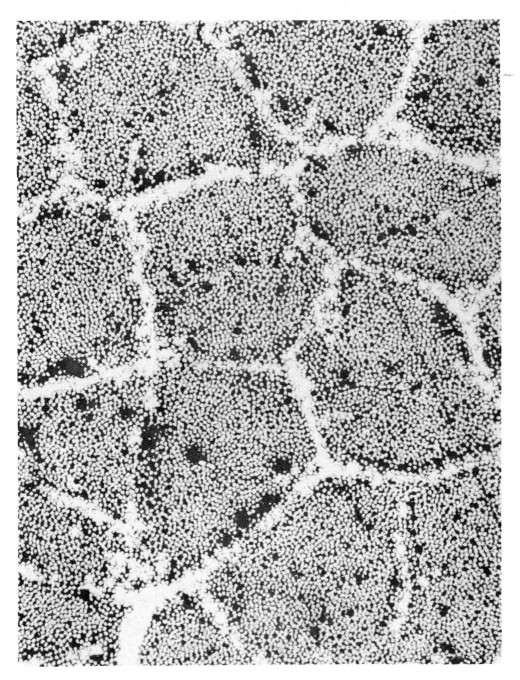

FIG. 10.8. Liquid infiltrated carbon–aluminium. (Kendall, E. G. (1974) *Composite Materials* **4,** 293, Kreider, K. G. (Ed.) Courtesy Academic Press.)

pressing are chosen to ensure adequate bonding within and between laminae without too much degradation of the fibres.

The sheet matrix method can be used to produce laminates directly.

10.2. Properties

As with reinforced polymers, a wide range of properties is possible with the variety of fibres available, choice of volume fraction, and choice of metal matrix.

Many reinforced metals examined so far have been model systems (these include sapphire–silver and tungsten–copper) fabricated to investigate the fundamentals of reinforcement, rather than with any application in mind. These will be by-passed here, and the systems that were investigated because they seemed likely to have some practical applications will be described. Considerable attention will be given to high-temperature properties, since this is the area in which reinforced metals are most likely to make themselves useful.

10.2.1. Strength and Modulus

With reinforced metals the metal normally yields before the fibre reaches its breaking-strain. Thus, in contrast to the polymer case, the metal can be expected to contribute most of its strength to the ultimate strength of the composite. (It does not usually contribute 100 % of its potential strength because the fibres have broken before it has fully work-hardened.) On these grounds, the composite may be expected to achieve a strength close to the Rule of Mixtures value. However, few do.

Table 10.1 gives some representative values for the fraction of the Rule of Mixtures strength that have been attained for reinforced metals with volume fractions close to 0.5.

The carbon–aluminium (the matrix is actually an aluminium–silicon alloy) achieves the high value because the silicon suppresses the reaction between carbon and aluminium. The tungsten–nickel alloy value is less than 1.0 because of internal stresses that are generated on cooling from room temperature; the efficiency of this composite increases as the temperature is raised, up to the temperature at which it was hot pressed. In the other cases the chemical reaction accounts for the low values. With *in situ* composites there is often some uncertainty about the "fibre" strength. However, Rule of Mixtures strengths have been obtained in some cases where the fibre strength could be determined.

Table 10.1 also gives values for the fraction of the Rule of Mixtures modulus that can be achieved in the fibre direction. Even when the strength is low, the fibres are still able to have a large stiffening effect. (Note, however, that silica and aluminium have almost the same modulus, so in this case there is no stiffening effect.) The elastic properties in directions oblique to the fibre direction are generally found to obey the theoretical relationships presented in Chapter 4, at least where the longitudinal modulus fraction is close to 1.0. With *in situ* composites, the moduli of the phases are often not known. The moduli of nickel- and chromium-based *in situ* composites range from about 140 to 300 GPa, as compared with a matrix modulus of 200 GPa for nickel and 279 GPa for chromium.

The best transverse strength that can be achieved is the strength of the unreinforced matrix. Table 10.1 shows that with two systems this can be achieved. With boron–

TABLE 10.1. *Reinforcement Efficiencies for Continuous Aligned Fibre Reinforced Metals.*
The values given are the fraction of the Rule of Mixtures values for strength and modulus, and the fraction of the matrix strength in the transverse direction.

System: ⎰ Matrix	Aluminium			Nickel Alloy		Titanium
⎱ Fibre	Boron	Carbon‡	Silica	Carbon	Tungsten	Boron
Longitudinal strength	0.70⁺	1.05	0.27	0.63	0.90	0.78
Longitudinal Modulus	0.97⁺	0.91	1.00	0.80	–	0.94
Transverse strength	1.00	0.25	1.00	–	–	0.60

⁺ Boron was coated with SiC (BORSIC).
‡ Matrix was 0.88 Al–0.12 Si.

aluminium the aluminium matrix fails, while with boron–titanium the boron fibres fail transversely to give the low value of 0.6. This transverse fibre failure is probably due to the high processing temperatures generating large residual stresses which, despite the subsequent annealing, weaken the fibres. The very low value of 0.25 for carbon–aluminium is the result of premature failure of the matrix caused by impurities introduced into the aluminium during the manufacturing process. This can probably be prevented by more careful control during fabrication. *In situ* composites should normally have the transverse strength equal to the matrix strength. However, when lamellar precipitates are obtained, the transverse strength in the plane of the lamellae can be very much greater than this, while that normal to the lamellae is then generally less.

It should be noted that the best transverse strengths are much greater than those obtained with reinforced polymers. For the boron–aluminium described in Table 10.1, various aluminium alloys were used as matrix, and the transverse strength obtained with the strongest alloy was 0.3 GPa. (Transverse strengths of aligned fibre reinforced polymers are typically 0.02 GPa.) However, this strength is much less than that in the fibre direction (1.8 GPa), and a good non-reinforced aluminium alloy can have a strength in all directions of 0.65 GPa (Table 1.1).

Stress–strain curves for moderate volume fraction (0.2 to 0.7) aligned continuous brittle fibre reinforced metals usually have two linear regions, the change in slope (the "knee") occurring at the matrix-yield strain. Near the breaking-stress there is sometimes a small pseudo-plastic region, during which the fibres are breaking into short lengths. Figure 10.9 shows the stress–strain curve for boron–aluminium, as fabricated, which shows these features. The aluminium matrix can be hardened by heat treatment; this roughly doubles the strain at the knee, increases the tensile strength to about 2 GPa, and suppresses the pseudo-plastic region. When the fibres are ductile, there is a true plastic region, as in the case of tungsten–nickel alloy shown in Fig. 10.10. The plastic region at the highest stress is due to the tungsten deforming plastically and is extensive for 300 C and 700 C. The usual knee, due to matrix yield, is not noticeable in this case. With *in situ* composites stress–strain curves resembling non-reinforced metals can be obtained, due to extensive plastic deformation of the fibrous or lamellar phase. An example is shown in Fig. 10.11; this material can be heat treated to increase the yield stress of the matrix.

Little work appears to have been carried out on the compressive strength of reinforced

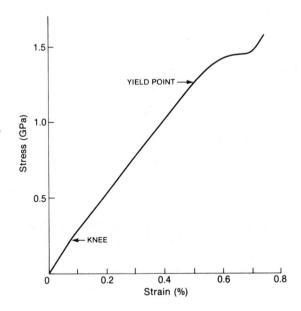

FIG. 10.9. Stress–strain curve for boron–aluminium. (After Kreider, K. G. and Prewo, K. M. (1974) *Composite Materials* **4**, 432, (ed. Kreider, K. G.), Academic Press, New York.)

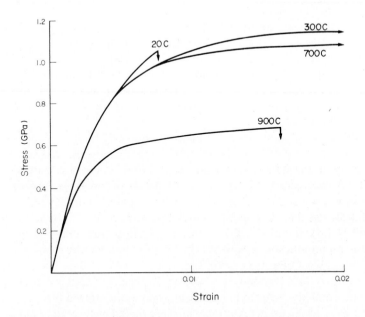

FIG. 10.10. Stress–strain curve for tungsten–nickel superalloy. (Courtesy of Westinghouse Canada Ltd.)

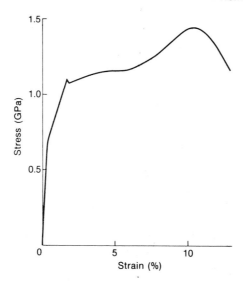

FIG. 10.11. Stress–strain curve for *in situ* composite (Ni–20Co–10Cr–3Al–TaC). Note the large plastic region. (After Bibring, H., Seibel, G., and Rabinvitch, M. (1972) *Mem. Sci. Rev. Met.* **69**, 41.)

metals. What results there are suggest that it is equal to or greater than the tensile strength. In the case of continuous aligned boron–aluminium, with $V_f = 0.6$, the compressive strength (2.14 GPa) was nearly twice the tensile strength in the fibre direction, and at 30°, 60°, and 90° to the fibre direction the compressive strength was greater than the tensile strength. In the case of carbon–nickel, the tensile and compressive strengths were the same (0.69 GPa) for $V_f = 0.5$ continuous aligned Thornel-75 graphite.

10.2.2. Toughness

The presence of the fibres usually reduces the toughness of the material. Despite this, in the case of B–Al, for fibres normal to the crack plane, the toughness increased with increasing volume fraction, in the range 0.3 to 0.5. Results from impact tests on a number of composites are given in Table 10.2, and compared with the impact toughness of the unreinforced matrix material.

The value for boron–aluminium is about one-third that of boron-epoxy (Table 9.4). The *in situ* alloys fare much better, the value for niobium carbide–nickel being more than twice that of Kevlar-epoxy. (Both the *in situ* composites included in the table have respectable tensile strenths—about 1.3 GPa—and Young's moduli in the fibre direction—about 300 GPa.)

The toughness of the composites is greatest when the fibres are normal to the crack plane (case A). When cracking occurs parallel to the fibres, the fibres in the crack plane split, or the interfaces fail at relatively low stresses, giving much lower works of fracture.

Increasing the temperature usually increases the work of fracture, due to the associated increase in the fibre toughness (especially with tungsten) and to the decrease in the matrix yield stress. The state of consolidation, and microstructure, of the matrix also has a large

TABLE 10.2. *Toughness Values, in kJ m^{-2}, Taken from Impact Tests at 20 C*

System		B–Al alloy (6061)	W–Ni[†] alloy	NbC–Ni[†]	(CoCr)–[†] (CrCo)$_7$C$_3$
V_f		0.5	0.6	0.11	0.3
Toughness		6[‡]	9[§]	340	94[¶]
		1.5[‖]	–	–	17[¶]
		1.5[‖]	–	–	14[¶]
	Matrix	130	240	630	–

[†] *In situ* composites.
[‡] Toughness increased with increasing V_f in range 0.3 to 0.5.
[§] Toughness increased to 100 kJm^{-2} at 370 C, and 500 kJm^{-2} at 1100 C. Hot working the material increased the 20 C toughness to 44 kJ m^{-2}.
[¶] Slow bending tests gave much lower values (approx. 1/10).
[‖] Toughness independent of V_f in the range 0.3 to 0.5.

effect on toughness; for example, hot working the tungsten–nickel alloy increased the toughness five-fold.

Fracture mechanics tests have been carried out on boron–aluminium. They have not, however, been successful so far, owing to the difficulty of obtaining sufficiently sharp notches (fatigue causes notch blunting rather than notch sharpening) and to the extremely large process zones at the tips of cracks in this material.

10.2.3. Fatigue

The fatigue resistance of fibre reinforced metals, like that of reinforced polymers, is excellent. Figure 10.12 shows the failure envelope for a boron–aluminium. This should be compared with the corresponding diagram for boron-epoxy, Fig. 9.17. (The fatigue

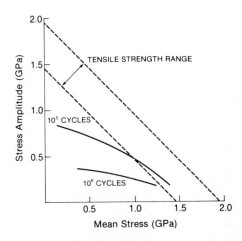

FIG. 10.12. Fatigue failure envelopes for boron–aluminium. (After Kreider, K. G. and Prewo, M. (1974) *Composite Materials* **4**, 459 (ed. Kreider, K. G.), Academic Press, New York.)

resistance is greater the closer the failure envelope is to the line joining the points for 1.6 GPa on ordinate and abscissae, 1.6 GPa being the tensile strength of this material.)

The fatigue resistance is mainly due to the excellent fatigue properties of the fibres, which are carrying most of the load. Excellent fatigue resistance is also found with *in situ* composites containing brittle fibres.

Interface failure is much less important with reinforced metals than with reinforced polymers, and failure is usually controlled by the metal matrix. This is because the matrix (especially aluminium) has relatively poor fatigue resistance.

In tension–tension fatigue the matrix can go into compression if its yield strain is exceeded during loading. The compression occurs on unloading, when the strain has decreased by an amount which exceeds the yield strain. Fibre failure can occur at stress raisers due to imperfections at the surface, or to internal matrix cracks caused by the alternating stresses.

It is considered that the excellent fatigue resistance of boron–aluminium, as compared with aluminium alloys may give impetus to its eventual use in lightweight structures.

10.2.4. High-Temperature Resistance

At least three problems have to be solved before composites can be used at high temperatures: (1) the components interact chemically and react with the environment; (2) differences in thermal expansion coefficients give rise to high internal stresses; (3) most materials creep at high temperatures.

Chemical interaction is not a problem with *in situ* composites. These are equilibrium systems so far as chemical potential is concerned. However, the large areas of interface results in high interfacial energies. Thus, at temperatures close to the melting-point, when diffusion rates become sufficiently great, the fibres tend to spheroidize (Fig. 10.13), leading to progressive loss in reinforcement. This process becomes serious when the temperature exceeds 0.9 T_m (T_m is the melting-temperature, K) and is assisted by stress gradients.

In the case of reinforced aluminiums, although the strength at temperatures up to 500 C can be quite good in short-term tests (e.g. for graphite–aluminium, 0.9 of its 20 C value) the strength falls with time at the temperature, and the maximum temperature for long-term use is much lower than this. For example, Fig. 10.14 shows that the strength of boron–aluminium falls rapidly at 540 C. Reinforced titanium will operate at higher temperatures; boron–titanium will survive about 10,000 hours at 540 C before it becomes critically weakened by titanium diboride formation at the fibre surface. This time is reduced to about four hours at 760 C.

With carbon–nickel, severe degradation of the carbon occurs at temperatures above 800 C due to the nickel promoting recrystallization of the carbon. In a cobalt matrix, the graphite recrystallizes rapidly at 700 C. Tungsten can be used to reinforce nickel at temperatures above 1000 C. The nickel promotes recrystallization and loss of strength of the tungsten fibres at high temperatures. The recrystallization is inhibited by hafnium carbide; at 1200 C no significant recrystallization occurs for at least 100 hours.

The interaction with the environment may usually be controlled by choice of matrix. For example, for high-temperature operation in the presence of combustion gases (in a turbine for example) nickel alloys are normally used because of their corrosion resistance.

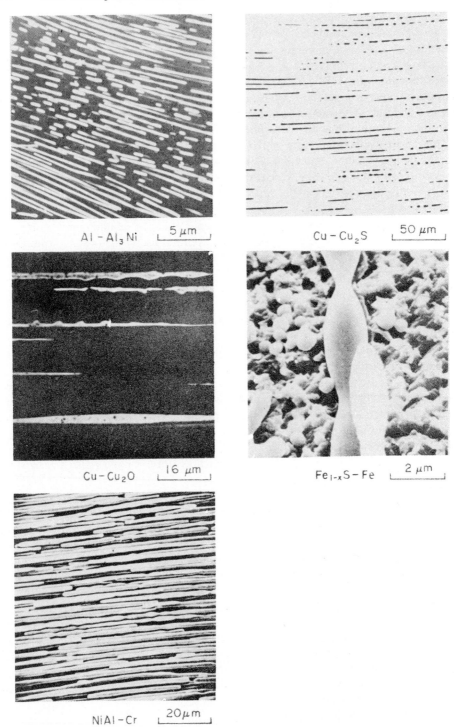

Al – Al$_3$Ni ⌊__5 µm__⌋

Cu – Cu$_2$S ⌊__50 µm__⌋

Cu – Cu$_2$O ⌊__16 µm__⌋

Fe$_{1-x}$S – Fe ⌊__2 µm__⌋

NiAl – Cr ⌊__20 µm__⌋

FIG. 10.13. Spheroidization in some *in situ* composites due to heating. (Weatherley, G. (1975) *Treatise on Materials Science & Technology*, **8**, 121 Courtesy of Academic Press).

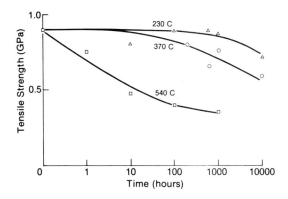

FIG. 10.14. Weakening of boron–aluminium due to heating. (After Sturke, W. F. Metal Matrix Composites, AIME, Symp. Pittsburgh 1969.)

At temperatures above 1000 C, however, oxygen readily diffuses through nickel, so oxidizable fibres have to be protected by diffusion barriers. Boron and carbon are both susceptible to attack by oxygen. Boron–titanium loses strength at 540 C due to the formation of boric oxide. (This oxide melts at 460 C.) Carbon in nickel above 600 C can oxidize rapidly and completely, to the gaseous state, leaving pores in the nickel. These results are summarized in Table 10.3.

TABLE 10.3. *Maximum Temperatures for Metals Reinforced with Boron, Carbon, and Tungsten Fibres*

System	Temperature (C)	Remarks
C–Al	500	Al contains 12 % Si
B–Al	540	B coated with SiC
B–Ti	650	B coated with SiC
B–Ti	540	Oxygen present; B coated
C–Ni	800	
C–Ni	600	Oxygen present
W–Ni	1200	W coated with HfC

Differences in thermal expansion between fibres and matrix can cause severe problems, and have rendered a number of potential reinforced metals unacceptable for high temperature use. Thermal expansion effects can give rise to very large stresses at the fibre–matrix interface. If the interfacial bond is good, these stresses can fracture the fibres, while if bonding is poor, complete separation can take place between fibres and matrix. Thus with short fibres or whiskers, reinforcement is lost. Thermal cycling is particularly damaging, and can lead to fibre break-up in many composites, including *in situ* composites. Figure 10.15 shows the longitudinal microstructure of Co, Cr–NbC after about 1500 cycles between 400 and 1120 C.

The thermal stresses can be calculated approximately if we consider the composite as made up of units, each consisting of one fibre surrounded by a tube of matrix, radius R_m.

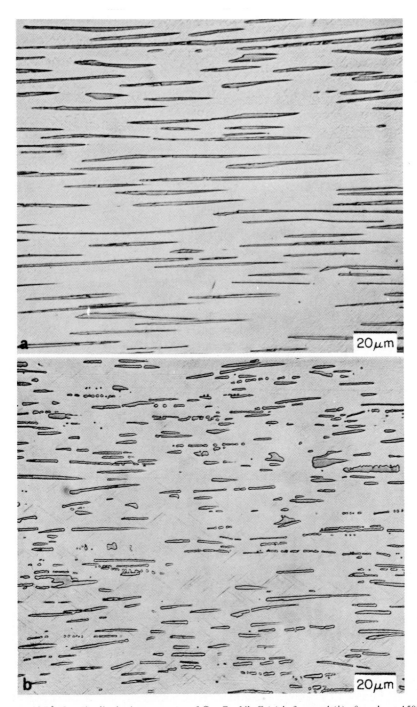

FIG. 10.15. Longitudinal microstructure of Co, Cr–Nb C (*a*) before and (*b*) after about 1500 cycles between 400 and 1120 C. (Thompson, E. R. and Lemky, F. D. (1974) *Composite Materials* **4**, 150 Kreider, K. G. (Ed.) Courtesy of Academic Press.)

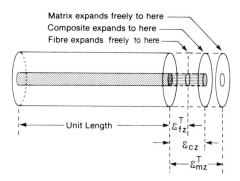

FIG. 10.16. Fibre, matrix, and composite axial displacements.

We consider continuous fibres, and Fig. 10.16 shows unit length of the composite, and the associated expansions.

The longitudinal displacements per unit length of "composite" are:

$$\varepsilon_{cz} = \varepsilon_{fz}^T + \varepsilon_{fz} = \varepsilon_{mz}^T + \varepsilon_{mz} \tag{10.1}$$

where ε_{fz}^T and ε_{mz}^T are the displacements of fibre and matrix if free to expand thermally, and ε_{fz} and ε_{mz} are the elastic strains of fibre and matrix, resulting from their need to expand the same amount when stuck together in the composite.

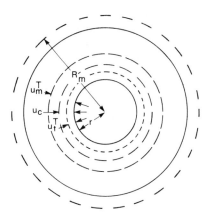

FIG. 10.17. Fibre, matrix, and composite radial displacements.

The corresponding radial displacements are shown in Fig. 10.17. We assume that fibre and matrix do not separate. Thus,

$$u_c = u_f^T + r\varepsilon_{fr} = u_m^T + u_m. \tag{10.2}$$

We assume uniform thermal expansion, so $u_f^T = r\varepsilon_{fz}^T$ and $u_m^T = r\varepsilon_{mz}^T$. ε_{fr} is the fibre elastic radial strain. u_m is the elastic matrix displacement, given in standard elasticity texts as $Pr(v_m + (1 + V_f)/V_m)/E_m$ where P is the internal pressure in the tube of matrix, resulting from the force exerted by the fibre expansion. V_f and V_m replace the r^2 and R_m^2 terms in the standard expression. (Note that R_m is different from R in equation 5.1; in this "composite" $V_f = r^2/R_m^2$.) In our case we must superpose a strain $-v_m \sigma_{mz}/E_m$ because of the axial stresses. Thus,

$$u_m = \{P[v_m + (1 + V_f)/V_m] - v_m \sigma_{mz}\}/r'E_m. \tag{10.3}$$

The stresses are

$$\sigma_{fr} = \sigma_{f\theta} = \sigma_{mr} = -P \tag{10.4}$$

where we are using the polar co-ordinates r and θ. Also,

$$\sigma_{m\theta} = P(1 + V_f)/V_m. \tag{10.5}$$

In the axial direction, equilibrium of the forces requires that

$$V_f \sigma_{fz} + V_m \sigma_{mz} = 0. \tag{10.6}$$

There are six stress–strain relations of the type

$$\varepsilon_z = \frac{1}{E}[\sigma_z - v(\sigma_\theta + \sigma_r)] \tag{10.7}$$

for the strains ε_{fz}, ε_{mz}, ε_{fr}, ε_{mr}, $\varepsilon_{f\theta}$, and $\varepsilon_{m\theta}$ (compare with equations (1.22) to (1.24)).

The fourteen simultaneous equations thus obtained can be solved to yield the stresses and strains. The resulting expressions are cumbersome, and can be simplified if

$$|v_f V_m E_m - v_m V_f E_f| \ll V_f E_f + V_m E_m. \tag{10.8}$$

This will usually be true for reinforced metals (and ceramics) so long as V_f is between about 0.2 and 0.5, the normal range for reinforcement. In this case we find that

$$P \simeq (\varepsilon_{fr}^T - \varepsilon_{mr}^T)V_m E_m E_f/[E_L + E_f/(1 + v_m V_m) - v_f E_m] \tag{10.9}$$

where

$$E_L = V_f E_f + V_m E_m.$$

Thus we have evaluated σ_{fr}, $\sigma_{f\theta}$ and σ_{mr}, while $\sigma_{m\theta}$ can be obtained from equations (10.9) and (10.5). Also

$$\sigma_{fz} = -P\{1 + [E_f(1 + v_m V_m + 2v_m V_f) - v_f E_m(V_m - V_f)]/E_L\} \tag{10.10}$$

and σ_{mz} can now also be determined, using equations (10.6), (10.9), and (10.10).

If α_f and α_m are the thermal expansion coefficients of fibre and matrix, then

$$(\varepsilon_{fr}^T - \varepsilon_{mr}^T) = (\alpha_m - \alpha_f)(-\delta T) \tag{10.11}$$

where $-\delta T$ is a temperature decrease, such as occurs during manufacture of the composite. If we examine the case for $V_f = 0.5$ we find that for many fibre reinforced metals equation (10.9) reduces to

$$P = 0.14 E_m(\varepsilon_{fr}^T - \varepsilon_{mr}^T) \tag{10.12}$$

with no more than about 5% error.

Table 10.4 gives some thermal expansion values, together with P values calculated from equation (10.12). These are quite large pressures, and since $\sigma_{mz} = -2P$ and $\sigma_{m\theta} = 3P$

TABLE 10.4. *Thermal Expansion Coefficients, and Thermally Induced Interfacial Pressures (P) for a Temperature Decrease of 500 C*
The expansion coefficients are the approximate values appropriate for the temperature range 0–1500 C in units of μK^{-1}. Composites have a fibre volume fraction of 0.5.
Note $\sigma_{fr} = \sigma_{mr} = \sigma_{f\theta} = -P$; $\sigma_{m\theta} = 3P$; $\sigma_{fz} = -\sigma_{mz} \simeq 2P$ for $V_f = 0.5$.

	Fibre	Boron	Carbon	Tungsten	NbC
Matrix	$\alpha \rightarrow$ $\downarrow$	8	8[†]	4	7
		Interfacial pressure (GPa)			
Nickel alloy	17	0.13	0.13	0.18	0.115
Aluminium	24	0.08	0.08	0.10	–
Titanium	8	0.00	0.00	0.03	–

[†] The values for carbon composites are only approximate since α_f (radial) = 8 and α_f (axial) = $0.5\,\mu K^{-1}$.

(while $\sigma_{mr} = -P$) there is a strong likelihood of a soft matrix like aluminium yielding under the high shears developed.

The stresses for the titanium matrix are lower than the others because of its lower thermal expansion coefficient. A more accurate analysis for C–Ti and B–Ti is indicated because $\alpha_f \simeq \alpha_m$, and we are neglecting the difference between the radial and thermal expansion of carbon, which will have an important effect in this case.

As with reinforced polymers, the creep resistance of fibre reinforced metals is generally excellent. Most of the systems likely to be of practical significance use fibres which have high melting-points, and the fibres are normally continuous. Thus creep under stress in the fibre direction is insignificant up to half the fibre absolute melting-temperature (K). Some nickel-based *in situ* composites have much better creep resistance than the superalloys at high temperatures, although the superalloys are better at lower temperatures. Figure 10.18 compares the creep resistance of Co, Cr–$(Cr, Co)_7 C_3$ and Ni–20Co–10Cr–3Al–TaC with a high strength nickel-based superalloy (TRW–NASA–6A) having exceptionally good properties. Both the composites are better than the superalloy at 1100 C for 1000 hours, but less good than it at 900 C. The advantages of the composites are greater when their lower densities are taken into account.

Tungsten–nickel superalloys also have superior creep strength (and creep strength/density) than the unreinforced alloys, and can be used in turbines at 25–50 C higher temperatures.

10.2.5. Joints

Joints are a problem with reinforced metals for the same reason that they are with reinforced polymers, i.e. because of the low shear strength of the material compared with its tensile strength. Thus large surface areas are required with joints. For example, with aligned boron–aluminium the tensile strength in the fibre direction is twenty times the shear strength in planes parallel to the fibres. Thus the joint length to component thickness ratio must be at least twenty, and is usually much larger than this because of stress concentrations at the end of the joint.

Butt and scarf joints cannot be used, and brazing fluxes are inadvisable because, with the

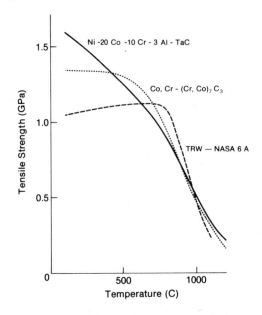

FIG. 10.18. Effect of temperature on the strength of two *in situ* composites compared with a nickel superalloy, TRW–WASA–6A. (After Thompson, E. R. and Lemky, F. D. (1974) *Composite Materials*, **4**, 136 (ed. Kreider, K. G.), Academic Press, New York, 1974).

large areas involved, some flux may remain trapped in the joint and weaken it. Thus fluxless brazing and solid state diffusion are the methods recommended. Both processes require great care to ensure that the fibres are not damaged by the heat and pressures used. With these processes joints having strengths approaching the matrix shear strength have been made. With boron–aluminium, for example, joint shear strengths of more than 50 MPa can be achieved. Joint design follows the general outline given for reinforced polymers in Section 9.4.

Further Reading

COTTRELL, A. H. (1968) *An Introduction to Metallurgy* (Edward Arnold, London).
KREIDER, K. G. (ed.) (1974) *Composite Materials 4, Metallic Matrix Composites* (Academic Press, New York).

11

Reinforced Ceramics, Cements, and Plasters

THESE matrices are characterized by extreme brittleness at normal ambient temperatures. Many are also very weak. Most of them are in a high state of oxidation, so are unreactive, even at high temperatures, but the presence of bound water in cements and plasters causes loss of integrity at elevated temperatures. These materials fall naturally into two groups: (1) high-temperature ceramics such as alumina and (2) materials that set as a result of hydration reactions at normal ambient temperatures, such as plasters, and calcium aluminate cements. (High-temperature metals, such as tungsten, which are often referred to as ceramic metals, will not be included here as matrices.)

Reinforced cements and plasters have a long history; the ancient Egyptians used straw to reinforce bricks. More recently, plaster reinforced with horse hair was used for internal finishing of buildings (e.g. lath and plaster ceilings) while concrete has been reinforced by relatively thick (1 cm) steel bars for more than a century.

Recent work on reinforcing these materials has had two main goals consistent with the two main matrix types. One objective is to produce a strong material for use at temperatures so high that metals or reinforced metals cannot perform adequately as load-bearing elements; i.e. above about 1200 C. This material could be used to construct more efficient engines, where its cost would be of secondary importance. The other aim is to produce very cheap materials which can take tensile as well as compressive loads at normal ambient temperatures. These materials are used in the construction of buildings, but if they could be made much stronger and more reliable, could be the main load-bearing elements for extended roofs, and bridges, etc.

Although the main reason for reinforcing the two different matrix types is usually the same, i.e. to increase the toughness, the differences in methods of manufacture and properties makes it appropriate that they be considered separately.

11.1. High Temperature Ceramics

11.1.1. The Matrix

High-temperature ceramics, such as alumina, are characterized by high modulus and medium strength (see Table 1.1) together with great hardness, extreme brittleness, and chemical inertness. They are usually made by casting the powdered material into shape with a fugitive binder. This "green" form is then hot pressed to produce an article with the required strength and density. It should also, if possible, have its final shape and size after hot pressing. This is because further shaping is very difficult, due to the hardness and brittleness of the material, and has to be done by fine grinding. Lower quality ceramics are fired without being pressed. This results in a higher void content than hot-pressing, and the final shape and size cannot be so closely controlled. However, it costs much less.

The firing or hot-pressing has to be carried out at very high temperatures, above 0.8 T_m (T_m is the absolute melting-temperature in Kelvins) in order to promote rapid diffusion and recrystallization, so that the densification proceeds to the required extent in a reasonable time. Thus temperatures of 1400 C are commonplace.

High-temperature ceramics are usually oxides, nitrides, or carbides of the lighter elements (e.g. Al, Si, Mg). However, another high-temperature ceramic is graphite, which is chemically reactive at high temperatures, and relatively weak and very soft, in contrast with the other ceramics. It also is made by hot-pressing and firing, but an inert atmosphere is needed.

11.1.2. The Interface

Fibre–matrix bonding is usually poor in reinforced ceramics. There are exceptional cases, however, in which chemical reactions occur which promote adhesion. With these, care has to be taken not to allow the reaction to proceed too far and damage the fibres. For example, at 1600 C zirconia fibres react with a magnesia matrix so that bonding is good, but if the temperature is raised to 1700 C the fibres are completely destroyed. (The zirconia migrates to the grain boundaries.)

When bonding is weak the thermal expansion of the components becomes very important. We showed, in Section 5.4, that the stress and strain at the slip point can be very small when adhesion is poor. Beyond the slip point stress transfer depends entirely on the frictional forces between the fibres and matrix in the case of reinforced ceramics. If the thermal expansion coefficient of the fibres is greater than that of the matrix there is a danger of the fibres separating from the matrix, so that the frictional force along the interface is zero. Thus we need fibres with relatively small thermal expansion coefficients.

The thermal expansion coefficients of the fibres must not be too small, however, since in this case large tensile stresses appear in the matrix parallel to the fibre axis, and cause it to crack.

We can extend the treatment in Section 10.2.4 to determine the maximum and minimum values for the fibre coefficient of thermal expansion when the fibres are not very rough.

Consider a composite element consisting of a single fibre surrounded by a tube of matrix (Fig. 10.16). Let the composite be under a strain ε_1, and consider the situation near the end of a fibre. Here, the axial fibre stress resulting from the stress transfer mechanisms discussed in Chapter 5 will be very small, and can be neglected for the moment. Thus our force equilibrium equation, equation (10.6), becomes,

$$V_f \sigma_{fz} + V_m \sigma_{mz} = V_m E_m \varepsilon_1. \tag{11.1}$$

The right-hand side is now $V_m E_m \varepsilon_1$ instead of zero. This affects the interfacial pressure, P, so that equation (10.9) becomes,

$$P = V_m E_f E_m [(\alpha_m - \alpha_f)(-\delta T) + \varepsilon_1 v_m] / [E_L + E_f (1 + V_m v_m) - v_f E_m]. \tag{11.2}$$

We only have positive pressure, and hence frictional stress transfer when,

$$\alpha_f < \alpha_m + \frac{v_m \varepsilon_1}{-\delta T}. \tag{11.3}$$

This gives a maximum value of α_f. Our minimum value is determined by matrix transverse cracking. We assume that for useful reinforcement the composite stress at the onset of matrix cracking σ_{1cr} should exceed the matrix strength, i.e. $\sigma_{1cr} > \sigma_{mu}$. Since we are dealing with brittle matrices this means that the associated strain should be elastic, i.e. $\varepsilon_{1cr} > \sigma_{mu}/E_L$. But the cracking-strain falls short of the matrix ultimate strain because of the extra internal strain due to differential thermal contractions, i.e. $\varepsilon_{1cr} = \varepsilon_{mu} - \varepsilon_{mz}$. Thus

$$\varepsilon_{mu} - \varepsilon_{mz} > \sigma_{mu}/E_L$$

and this can be rearranged to give

$$\varepsilon_{mz} < V_f \sigma_{mu}(E_f - E_m)/E_m E_L \tag{11.4}$$

where we calculate ε_{mz} in the absence of ε_1. One of equations (10.7) is

$$\varepsilon_{mz} = [\sigma_{mz} - v_m(\sigma_{m\theta} + \sigma_{mr})]/E_m.$$

Substitution of σ_{mz} from equations (10.10) and (10.6), $\sigma_{m\theta}$ from equation (10.5), and σ_{mr} from equation (10.4) and rearranging gives,

$$\varepsilon_{mz} = \frac{PV_f}{V_m E_m}\{1 + [E_f(1 + v_m V_m) + E_m(v_f V_m - v_f V_f - 2v_m V_m)]/E_m\}. \tag{11.5}$$

When we substitute inequality (11.4) for ε_{mz}, and substitute for P from equation (10.9) we obtain the inequality for α_f

$$\alpha_f > \alpha_m - \frac{\varepsilon_{mu}}{-\delta T}(1 - E_m/E_f)E_{mf} \tag{11.6}$$

where

$$E_{mf} = \frac{1 + [E_f(1 + v_m V_m) - v_f E_m]/E_L}{1 + [E_f(1 + v_m V_m) - E_m(v_f V_m - v_f V_f - 2v_m V_m)]/E_L}.$$

This can be simplified since $E_{mf} \simeq 1$ ($E_{mf} = 1$ for $v_f = v_m$) and we can combine both our conditions (11.3) and (11.6) in one expression,

$$\alpha_m - \frac{\varepsilon_{mu}}{-\delta T}(1 - E_m/E_f) < \alpha_f < \alpha_m + \frac{v_m \varepsilon_{mu}}{-\delta T} \tag{11.7}$$

(Note that we have replaced ε_1 in inequality 11.3 by ε_{mu}, this being nearly the highest value ε_1 can take before matrix cracking occurs. It can be a little higher than ε_{mu} because, when α_m is near the upper limit, the matrix is under longitudinal compression).

Relevant data for three matrices are given in Table 11.1. From these data the serviceable range for α_f, for good room temperature properties, has been calculated using expression (11.7). These results are also given in the table. Apart from Pyrex glass, which is not a particularly high-temperature matrix, the range for α_f is very small.

Table 3.3 lists the expansion coefficients of fibres. It is clear why there are difficulties in

TABLE 11.1. *Matrix Data, and Serviceable Range for Coefficients of Expansion for Reinforcing Fibres with Smooth Surfaces*
(Poisson's ratios for fibres and matrices will be found in Table 6.1).

Matrix	T_c (C)	σ_{mu} (Gpa)	E_m (GPa)	α_m (μK^{-1})	$\alpha_f min^\dagger$ (μK^{-1})	$\alpha_f max^\dagger$ (μK^{-1})
Pyrex glass	520	0.10	60	3.5	0.8	4.2
Glass ceramic‡	1000	0.10	100	1.5	0.8	1.7
Alumina	1400	0.28	400	8.8	8.8	8.9
Silicon	600	0.45	161	4.1	3.8	4.2
Silicon nitride	1450¶	0.3	307	2.87‖	2.7	3.0§

† Minimum and maximum values for fibre thermal expansion coefficient.
‡ Lithia alumino silicate.
§ Calculated assuming $v_m = 0.2$.
¶ Nitriding temperature.
‖ α-Si$_3$N$_4$, α_m for β-Si$_3$N$_4$ = 2.25. The corresponding α_f's are 2.1 and 2.4.

getting reinforcement with such high-temperature fibres as C, SiC, Mo, and W, since the coefficients generally lie outside the appropriate ranges.

With unfavourable contractions, reinforcement can still be obtained, if the fibres are roughened to match the relative contractions. The fibres should be etched or otherwise modified to produce corrugations of depth h where,

$$h/r > (\alpha_f - \alpha_m)(-\delta T) + v_m \varepsilon_{mu}. \tag{11.8}$$

(This comes from inequality (11.3) with $\varepsilon_1 = \varepsilon_{mu}$ and h chosen to exceed the total displacement.) The roughness need not be very great; for example for SiC–Si$_3$N$_4$ we find that h only needs to be about $0.003r$, or a depth of about $0.15\ \mu m$ for normal diameter silicon carbide fibres.

The roughening only helps with the upper limit for α_f, however, and care must be taken that the fibres are not unduly weakened by the roughening process, and that pull-out is not inhibited significantly, so that the fracture toughness is not impaired.

11.1.3. Methods of Manufacture

The same methods used for the manufacture of unreinforced ceramics may be used to make reinforced ceramics. Hot pressing is used for both random and aligned fibre composites. In the random fibre case, the chopped fibres may simply be mixed with the powered matrix prior to hot-pressing. With aligned fibres a "prepeg" tape is made by, for example, extruding the fibre–matrix mixture with a binder, such as ammonium alginate. (This binder hardens in an acid bath, so the extrusion is done into such a bath.) Aligned continuous fibre tapes may be made by passing the fibres through a mixture of powdered matrix and binder. These tapes are then cut into suitable lengths and hot pressed.

The hot-pressing process can produce parts which are almost completely free of voids. The temperature typically exceeds 1200 C. The dies are often made of graphite, since this has a low coefficient of thermal expansion, and the part can easily be removed from the mould after cooling. However, graphite dies are brittle, are easily abraded, and are rapidly oxidized in air at these temperatures. They should be used in an inert atmosphere, and do not last long. Alternatives include molybdenum–titanium–zirconium alloys, and silicon carbide, but these have higher expansion coefficients, and other problems.

Hot-pressing is time-consuming and expensive. Pressing can be carried out more economically at lower temperatures in a two-step process with some loss in mechanical properties. The first step is to press at relatively low temperatures to partly consolidate the material and give it the desired shape. The second step is to heat the partly consolidated parts at a very high temperature without pressure, to promote further densification. The density of the parts produced is lower than can be obtained by hot-pressing, and the shape is much less well controlled, since the parts can warp while unrestrained at high temperature, and contraction may not be uniform.

Fibres can be incorporated in glassy ceramics by high-temperature casting. This can only be done with relatively low fibre volume fractions (usually < 0.1) since high volume fractions increase the viscosity of the mixture too much. Extrusion may also be used, but the fibre volume fraction is still rather restricted.

Carbon reinforced carbons are made by pyrolizing carbon reinforced polymers (e.g. phenolics or epoxies), and then graphitizing the matrix by further heating. They are also made by coating woven fibre forms with pitch, and this is pyrolized in a hot isostatic press, and subsequently graphitized; several impregnation and pyrolization cycles are needed to complete the densification process before graphitization is carried out. The pyrolization step is carried out under high pressure in order to enhance the yield of carbon from the pitch. About 85% conversion can be obtained at 100 MPa, while only about 50% can be achieved at 0.1 MPa. The pyrolization takes place at 550–650 C and takes about 24 hours per cycle. Graphitization is carried out at about 2700 C for about one hour. Figure 11.1 illustrates the steps in this process. Carbon–carbons can be made with three-dimensional reinforcement using continuous fibres. This is done by special weaving processes which will produce sheets 2 cm or more in thickness.

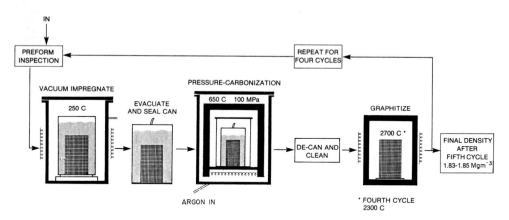

FIG. 11.1. Impregnation and densification of carbon–carbons. (Courtesy of Fiber Materials Inc.)

In situ composites can be made by controlled crystallization of molten ceramic alloys (for a discussion of this method, see Section 10.1.3). The reinforcement can be in platelet or fibre form, and refractory compounds as well as elements are suitable. Tungsten reinforced zirconia, hafnia, and urania have been produced, also Mo–Gd_2O_3/GeO, and plates of $BaFe_2O_4$ in $BaFe_{12}O_{19}$. In addition chromium and molydenum fibres can be grown in Cr_2O_3 and $(Cr, Al)_2O_3$; an example is shown in Fig. 11.2.

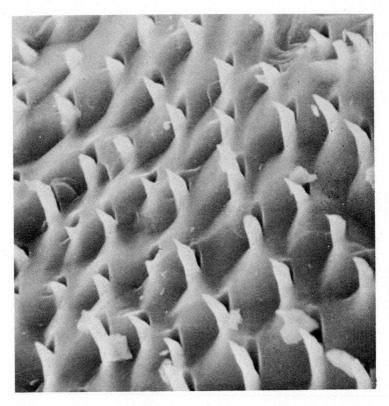

FIG. 11.2. Mo fibres grown in $(Cr, Al)_2O_2$. (Courtesy of Nils Claussen, Max Planck Inst.)

Some of the most interesting reinforced ceramics have been made by reacting carbon fibres with silicon to produce *in situ* composites with SiC fibres. The fibres are infiltrated by molten silicon at 1450–1600 C and kept at temperature for up to 5 minutes to complete the reaction. Excess silicon forms the matrix. The fibres are extremely rough, providing very good mechanical interaction, but are not very strong ($\simeq 0.5$ GPa) (Fig. 11.3).

The matrix also may be produced by *in situ* reaction. Silicon nitride matrix composites have been made by flame spraying (and plasma spraying) silicon onto a fibre form, then nitriding the silicon by heating in nitrogen to 1450 C. A carefully controlled warm-up is required, and the whole nitriding process takes over 5 days. The nitride formation is accompanied by an increase in volume, which results in densification of the product. Fibres used for this process have to be resistant to the high temperatures involved, and unreactive. Silicon carbide, and tungsten or carbon coated with silicon carbide are suitable. Silicon doughs with fugitive binders are also suitable for the production of these composites.

11.1.4. Strength and Modulus

Continuous aligned fibre-reinforced ceramics usually have strengths and moduli which are somewhat below the values that are theoretically possible, according to the Rule of

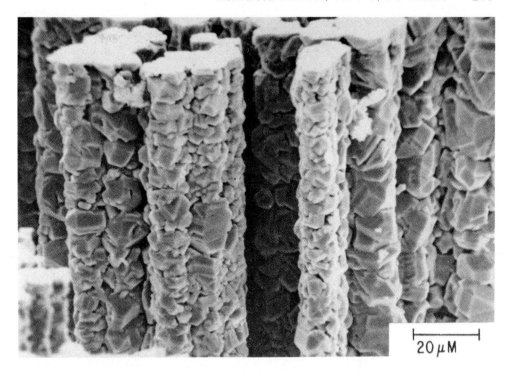

20 μM

Fig. 11.3. SiC fibres growth from carbon during liquid infiltration by silicon. (Mehan, R. L.
(1978) *J. Mater Sci.* **13,** 358 Courtesy of Chapman & Hall.)

Mixtures. In this respect they compare unfavourably with reinforced polymers and metals, where agreement with the mixture rule can be obtained, so long as sufficient care is taken with manufacture.

Table 11.2 shows some typical values for the fraction of the Rule of Mixtures strength and modulus that have been obtained with aligned continuous carbon fibre reinforced ceramics. Of particular note is that the modulus fraction is less than 1.0, except in the case of the carbon–carbon. These low results show that the matrix is not transferring the stress efficiently to the fibres, and the reason for this is almost certainly the separation of fibres and matrix due to the thermal contractions after manufacture. Smooth fibres can then slide in their holes in the matrix, hampered (and stressed) only by adventitious circumstances such as kinks and curvature in the holes.

The true modulus fraction could be much lower than indicated in the table, since the sample were tested in flexure. This test conceals any change in modulus with increase in stress, since most of the specimen is at a stress which is much lower than the maximum. (See Section 1.2.1 for the stress distribution in flexure.) Note that in the case of ceramics (and other brittle materials where failure is governed by surface imperfections) the flexure test normally overestimates the strength.

The carbon–carbon can achieve the Rule of Mixtures modulus because the graphite matrix has about the same coefficient of expansion as that of the fibres in the radial direction. SiC–Si also achieves the Rule of Mixtures modulus, probably because of the excellent keying of the very rough fibres.

TABLE 11.2. *Fraction of the Rule of Mixtures Values Obtained for Strength and Modulus of Carbon Fibre Reinforced Ceramics Fibres are continuous and aligned, with V_f between 0.4 and 0.5*

Matrix	Modulus fraction	Strength fraction
Pyrex glass	0.86	0.81
Glass ceramic	0.71–0.80	0.81
Carbon	1.00	0.50

The strength fractions shown in Table 11.2 are not very different from the modulus fractions. Although the heat and pressure used in manufacture of the composites can cause a loss in fibre strength, these results were chosen to illustrate what can be achieved when this problem is avoided.

The strength of aligned continuous fibre ceramics increases with volume fraction, and this increase is often linear up to $V_f = 0.6$. Figure 11.4 shows two examples. Carbon–Pyrex follows a Rule of Mixtures expression, with an effective fibre strength of about 0.78 GPa. The carbon–lithium aluminosilicate does not follow the Rule of Mixtures, since it extrapolates to negative values of σ_{1u} at $V_f = 0$.

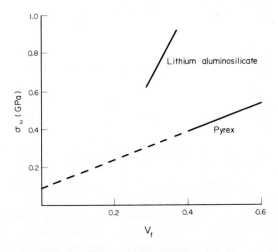

FIG. 11.4. Flexural strength parallel to the fibres in aligned continuous carbon–Pyrex and carbon–lithium aluminosilicate. (After Levitt, S. R. (1973) *J. Mater. Sci.* **8**, 793, and Sambell, R. A. J., Bowen, D. H. and Phillips, D. C. (1972) ibid. **7**, 663.)

The strength of SiC–Si follows a Rule of Mixtures expression with an effective fibre strength of 0.5 GPa. Very high volume fractions can be achieved in this case, and the properties depend strongly on the good fibre–matrix keying. However, this is achieved at the cost of having relatively weak fibres. Good keying is necessary because α_f for SiC is above the upper bound for reinforced silicon (see Tables 11.1 and 3.3). This material also has good transverse strength, as is shown in Fig. 11.5. The high value of σ_{2u} indicates a different mode of failure from that considered in Section 6.8. It is possible that the fibres

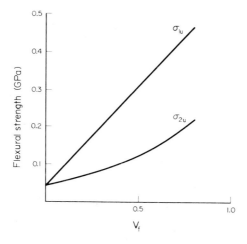

FIG. 11.5. Strength parallel (σ_{1u}) and normal to the fibres (σ_{2u}) in aligned continuous SiC–Si. (After Mehan, R. L. (1978) *J. Mater. Sci.* **13**, 358.

form a three-dimensional network when they are converted from C to SiC, with most of the strength in the original carbon fibre direction.

When the fibres are randomly oriented, the strength often decreases monotonically with increasing V_f. Figure 11.6 shows some examples. The effect is less severe if the thermal mismatch between fibres and matrix is small, and increases in strength are obtained when it is zero (the Al_2O_3 – porcelain and Al_2O_3 – glass in Fig. 11.6). As shown in Section 11.1.2 it is not advantageous to have a matrix with a much larger thermal contraction than the fibres, because of the risk of transverse matrix cracking.

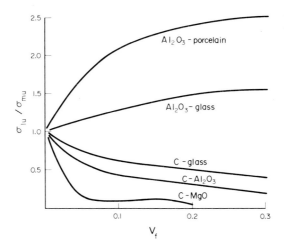

FIG. 11.6. Normalized flexural strength of randomly oriented short brittle fibre ceramics. (After Donald, I. W. and McMillan, P. W. (1976) *J. Matter. Sci.* **11**, 949.)

Randomly oriented soft, ductile fibres, e.g. Mo or Ni, at low volume fractions, can cause a decrease in the strength of the material. (They are put in to increase the toughness rather than the strength.) However, when V_f is increased to 0.1 to 0.2 the strength of the composite increases to about the matrix value, Fig. 11.7. Randomly oriented tungsten fibres increase the strength of glass and silica, and the effect increases with increasing V_f, at least at low volume fractions when $\alpha_m < \alpha_f$.

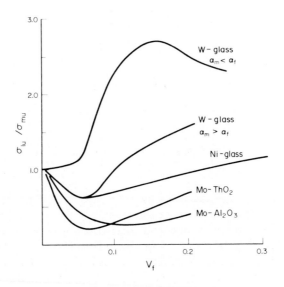

FIG. 11.7. Normalized flexural strength of randomly oriented short ductile fibre ceramics. (After Donald, I. W. and McMillan, P. W. (1976) *J. Mater. Sci.* **11**, 949.)

The shear strength of well-made aligned fibre reinforced ceramics is approximately equal to that of the unreinforced matrix. Thus, the discontinuities caused by the presence of the fibres do not constitute flaws which are any more serious than those which occur in the matrix anyway.

11.1.5. Toughness

The main advantage of putting fibres into ceramics is the increase in toughness obtained thereby. Brittle and ductile fibres both increase the toughness, and the increase is usually very large, and greater for larger volume fractions of fibres. At low fibre volume fractions the randomly oriented short ductile fibres (molybdenum and nickel) have the largest effects, but for volume fractions near 0.5 aligned fibres have to be used, and brittle fibres at these concentrations can give increases in toughness of many thousand-fold over the unreinforced matrix value. Figure 11.8 illustrates some of the results obtained. Slow bend test values of 6 kJ m^{-2} can be obtained with reinforced ceramics. These results should be compared with 40 kJ m^{-2} obtained with the same method with carbon-epoxy (Table 9.4), and the 3 Jm^{-2} typical of unreinforced ceramics. Aligned SiC–Si is not very tough. The fracture toughness varies linearly with V_f from 0.9 MPa m$^{-\frac{1}{2}}$ at $V_f = 0$ to 4.0 at

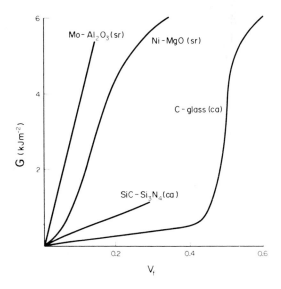

FIG. 11.8. Work of fracture of short random (sr) and continuous aligned (ca) fibre ceramics. (After Donald, I. W. and McMillan, P. W. (1976) *J. Mater. Sci.* **11**, 949.)

$V_f = 0.7$. (The corresponding **G** values are about 6 and 45 Jm^{-2}.) The low toughness is probably because the good mechanical keying causes fibre fracture rather than fibre pull-out.

11.1.6. Other Properties

Increase in temperature might be expected to improve the properties of fibre ceramics because the thermal expansion mismatch should decrease. However, data on SiC–Si shows a slight decrease in strength in the fibre direction as temperature is increased, with a larger decrease as the melting-point of silicon is approached (Fig. 11.9). The transverse strength increases at about 600 C, when the silicon becomes ductile, and increases even more at about 1000 C where the load-deflection curve becomes non-linear. It falls off again at about 1200 C (Fig. 11.9). When carbon fibre felts are used instead of fibre tows, strength is retained to above the melting-point of silicon. The modulus of the material also falls slightly as the temperature is raised.

The maximum temperature for the composite is often limited by oxidation or chemical interactions. Ceramic matrices are prone to developing microcracks, and in the case of tungsten and molybdenum reinforced systems, microcracked specimens have their fibres rapidly oxidized above 700 C. Reinforced carbons, and carbon ceramics must be used in non-oxidizing atmospheres above about 400 C, but under the appropriate inert conditions carbon–carbon can be used up to at least 2000 C. The non-reactive silicon carbide silicon can be used up to 1370 C in oxidizing atmospheres, while carbon lithium alumino silicates can be used at 1200 C in an inert atmosphere. Table 11.3 gives some data for theoretical maximum temperatures based on melting- or softening-points for a number of reinforced ceramics.

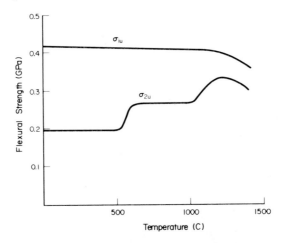

FIG. 11.9. Effect of temperature on the strength of SiC–Si. (After Mehan, R. L. (1977) *Amer. Ceram. Soc. Bull.* **56**, 211.)

TABLE 11.3. *Theoretical Maximum Temperatures for Some Reinforced Ceramics (Based on Melting or Softening-Points)*.

System	Maximum Temperature (C)
Carbon–Pyrex glass	700–800
Carbon–glass ceramic	1300
Silicon carbide–silicon	1410
Silicon carbide–silicon nitride	1900
Carbon–carbon	3550

Note: Carbon oxidizes in air above 400 C, and SiC oxidizes rapidly between 980 and 1150 C but is stable between 1150 and 1400 C.

A problem with the high-temperature application of reinforced ceramics is that thermal cycling can lead to destruction of the fibres or severe cracking of the matrix, or complete failure of the fibre–matrix bond. The magnitude of the effect depends on the mismatch of thermal expansion coefficients between fibres and matrix. The greater it is, the less thermal cycling the composite can withstand without severe loss of properties. Even carbon–carbon can suffer from this problem, due to the anisotropy of the thermal expansion of the fibres.

Cyclic stressing of reinforced ceramics in flexure indicates that irreversible changes occur on the first cycle. The flexure force-distance curve of many reinforced ceramics has a slight bend between 30 and 50% of the ultimate load. This bend disappears after the first cycle (Fig. 11.10). Thereafter, the force-distance curve shows slight hysteresis, and the slope of the curve is reduced by about 6%. (Note that this indicates a decrease in Young's modulus, but tensile tests are required to determine the extent of the decrease, which undoubtedly is much greater than 6%.) Cyclic stressing up to 10 cycles at 0.55 GPa did not reduce the strength of the material. This is to be expected with ceramic reinforced

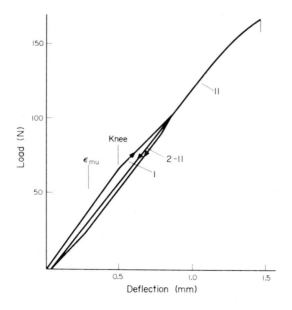

FIG. 11.10. Cyclic flexure of carbon glass ceramic. Labels on curves indicate cycle numbers.
(After Levitt, S. R. (1973) *J. Mater. Sci.* **8**, 793.)

ceramics: usually neither component is weakened by dynamic fatigue, so that as long as the interface does not lose its ability to transfer stress, the composite will be highly fatigue resistant.

11.2. Reinforced Cements and Plasters

11.2.1. The Matrices

Portland cement is one of the most heavily used materials in the building industry. It is usually combined with aggregates (sand and stones) to make concrete. This is a weak material in tension, but is widely used to support compressive loads. For tensile load-bearing the concrete must be reinforced; steel bars are normally used for this, and bear almost all the tensile load.

Cement is made by heating chalk or limestone with clay, and consists of a combination of calcium oxide, aluminium oxide, and silica with a preponderance of the calcium oxide. The standard Portland cement composition is often referred to as tricalcium aluminate. Different compositions are available to suit different applications. Cement paste is the mixture of cement with water. In the presence of water, interlocking and adhering crystals of hydrated calcium alumino silicates grow to produce a hard solid. The setting process is slow, and although Portland cement begins to harden after a few hours, it requires at least 24 hours to achieve good cohesion, and 28 days are usually allowed to elapse before samples of the material are tested for strength, etc. The hardening process continues indefinitely, but high-alumina cements set faster than Portland cement. The best results are obtained when the setting or curing process takes place in an atmosphere with 100%

relative humidity, or with water completely surrounding the mould. Shrinkage occurs during curing, and, in addition, the material will creep under load.

The cement structure is full of flaws, and very weak. For the hardened cement paste the strength is typically 7 MPa, while for mortar with 50 % sand it is about 5.5 MPa, and for concrete (this includes stones) it is about 3.5 MPa. The strength and modulus are greatly affected by the water/cement ratio. The presence of aggregates increases the stiffness: the moduli are respectively 20, 30, and 50 GPa for cement paste, the mortar, and concrete.

Plasters are generally not used in load-bearing situations on account of their weakness. They were first developed as materials for finishing rough brick, stone, or concrete surfaces. More recently they have been made in the form of sandwich sheets, with paper on the outside. This makes the material sufficiently strong to the self-supporting over short spans (less than 1 m), so that the sheets can be attached directly to wooden ceiling joists and wall studs, reducing the labour involved in constructing ceilings and walls in private dwellings. There are two main types of plaster: lime and gypsum. Lime plaster is a soft mixture of calcium hydroxide and water. After being applied to a surface, it is oxidized by carbon dioxide in the air to form a hard calcium carbonate layer. This resists further absorption of carbon dioxide. It must be applied in very thin layers. Usually three layers or more are applied, with intervals between the applications to allow for hardening. The material shrinks as it dries, causing voids. Sand is often added to reduce the proportion of water required, so reducing shrinkage.

Gypsum plaster, or plaster of paris, is made by dehydrating gypsum; $CaSO_4.H_2O$ is converted to $CaSO_4.\frac{1}{2}H_2O$ by heating to 120 C. When mixed with water the powdered gypsum plaster hardens by reverting to gypsum with the growth of long thin interlocking crystals. Plasters are generally weaker than cement, and gypsum typically has a strength of about 4 MPa. A slight increase in volume occurs during setting, and the hardening takes place very quickly, a few minutes only being required. Sandwich sheets are usually made with gypsum plaster, faced with paper, and are often referred to as gypsum board.

11.2.2. The Interface

Cement is strongly alkaline and if well made is only very slightly water permeable. Thus it provides a protective environment for steel. When thin steel wires or fibres are used, some rusting occurs during the early stages of the curing, when free water is still present. This normally does not cause a significant loss in strength. Most glass fibres are severely corroded in the alkaline environment. Thus they have to be protected, or special glasses have to be used. Early glass reinforced concretes were made with polymer coated glass fibre bundles, but these are giving way to reinforcement by special glasses, which do not need to be coated. Figure 11.11 compares the corrosion of E-glass with one of the special alkali resistant glasses.

An adequate bond is obtained between cement and fibres by virtue of the shrinkage that takes place on curing. In addition there is usually some mechanical keying due to surface roughness in the case of steel (this is enhanced in reinforced concrete by using heavily indented steel bars). Thus reinforced cements are usually made without using special coatings on the fibre surface.

With glass reinforced gypsum the glass is coated with polyvinyl acetate before being mixed with the gypsum slurry. This gives a good interfacial shear stress, varying from

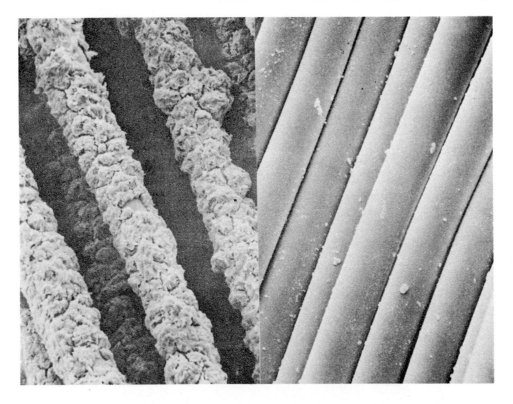

FIG. 11.11. (a) Normal glass fibres, (b) alkali-resistant glass fibres after exposure to cement environment. (Courtesy of Fibreglass UK Ltd.)

1 MPa for poorly compacted composites to 7 MPa for well-compacted ones. The interfacial shears seem to be entirely frictional.

11.2.3. Methods of Manufacture

Some of the methods already described for making reinforced thermoset resins are suitable for making reinforced cements and plasters. Since curing takes place at room temperature, the process is simplified by the omission of the heating needed for curing most thermosets.

The method of construction used depends on fibre length. With very short fibres, large mouldings can be made using normal concrete placing methods, i.e. simply pouring the mixture into the mould, and vibrating to consolidate the material and promote the escape of trapped air. The amount of fibre that can be added is severely limited (usually less than 5 %) because the fibres increase the viscosity of the mixture, and inhibit consolidation.

Longer fibres can be used, and higher volume fractions achieved by hand lay-up (see Section 9.1.4) and spray-on (Section 9.1.5) methods. Figure 11.12 illustrates the spray-on process for glass cement. The fibre bundles retain a great deal of their integrity as bundles in this process. The cement mixture used often contains excess water to facilitate impregnation of the fibres and reduce the amount of air voids. This must be removed by

FIG. 11.12. Spray-suction method for the production of glass cement.
(Courtesy of Fibreglass UK Ltd.)

suction from behind the preform screen. In the hand lay-up process fibre cloths and tapes impregnated with a cement slurry may be used.

Fibre cloths and tapes may also be wound, as in filament winding (see Section 9.1.6). The material first passes through a bath containing the cement slurry (or cement/polymer latex slurry—the addition of small amounts of suitable polymers increases the breaking-strain of the matrix) and is then wound on a mandrel. Figure 11.13 shows a typical section of pipe made by this method.

Panels are produced from fibre cloths by passing the material through the slurry, then through rollers to ensure complete infiltration, and squeeze out excess matrix. Continuous sheets are produced, which are then cut to size.

The tape may be made more hydrophilic by the inclusion of a small proportion of water-absorbing fibre (e.g. cellulose). This greatly improves the take-up of cement from the bath, and assists impregnation.

FIG. 11.13. Pipe made from glass-cement by tape winding.
(Courtesy of Dempster, D. P., Government of Ontario.)

Asbestos cement is made in a way which is analogous to the sheet-moulding compound (Section 9.1.3). The asbestos can be dispersed relatively easily in water, and a dilute solution (6 % solids) of the asbestos and cement is first made. The solids are picked up from this as a thin film on a rotating drum of wire mesh. They are then transferred to a felt screen on which they are partly dried by suction. Excess water is removed and final compaction takes place on another steel drum, on which the layers are collected until the desired thickness is produced. This is then cut off to make flat sheets which can be shaped and moulded as desired. Asbestos cement pipes are made in a similar fashion.

11.2.4. Strength and Modulus

Great improvements can be made in the strength of cements by the addition of small amounts of asbestos, glass, steel, kevlar, or carbon fibres, since the cement alone is so weak.

Well-made continuous aligned fibre cements obey the Rule of Mixtures quite closely for modulus at low applied stresses, and for strength.

In the case of carbon, very little has been reported on random fibre composites. With aligned carbon fibres V_f should not exceed 0.12 since poor compaction is obtained. This is due to the relatively large size of the cement particles (they can be a lot bigger than the fibre diameter).

With the other fibres most work has been done with randomly directed fibres in cement sheets. With these composites good compaction can only be obtained at low V_f. With asbestos V_f should not exceed about 0.12, while for other fibres it should not exceed about 0.08, the maximum depending on fibre aspect ratio.

The flexural and tensile strengths (Fig. 11.14) are a linear function of V_f up to the maximum volume fraction for good compaction (0.06 in the case of the glass cement shown in Fig. 11.14). The strength is usually greater for the longer fibres, as shown in Fig. 11.15. However, good compaction is difficult with long fibres, and best results are obtained when V_f s has some maximum value, as shown in Fig. 11.16 for steel.

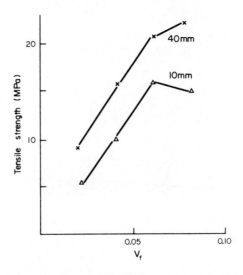

FIG. 11.14. Tensile strength of air-cured random glass cement.
(After Majumdar, A. J. (1975) RILEM, 279.)

The flexural strength is quite close to the theoretical (i.e. about $3\,V_f\sigma_{fu}/8$; see equation (4.57)). The tensile strength is very much less, about one-third of the flexural strength. Table 11.4 gives some typical values. Glass fibres do not affect the compressive strength significantly when $V_f = 0.06$, but steel gives a small increase, and asbestos a bigger one (Table 11.4).

Although some increase in modulus is obtained with aligned fibres (especially carbon), random fibres at useful volume fractions have negligible effect. The stress–strain curves show the features described in the theoretical section (Section 6.3). Aligned fibres have a plateau region (Fig. 11.17) where multiple cracking is taking place. (The "plateau" is not quite parallel with the strain axis because, after each crack has split the matrix, the remaining pieces of matrix are stronger than the original, larger piece.) The plateau terminates when the matrix has broken up into pieces of the critical size, and thereafter the slope of the stress–strain curve is equal to $V_f E_f$.

With random fibres, similar effects are observed (Fig. 11.18). The high breaking-strain of

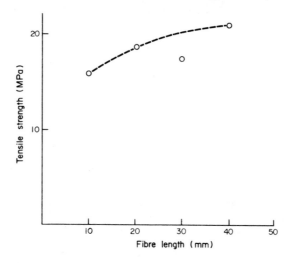

FIG. 11.15. Tensile strength of air cured random glass-cement.
(After Majumdar, A. J. (1975) RILEM, 279.)

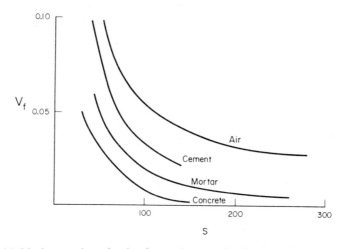

FIG. 11.16. Maximum volume fraction for good compaction for three-dimensionally packed
fibres of aspect ratio, s. (After Krenchell, H., A.C.I., SP44, p. 45.)

TABLE 11.4. *Typical Data for Reinforced Cements 28 days after Manufacture*

| Fibre | V_f (%) | Strengths (MPa) | | | Toughness $kJ\,m^{-2}$ |
		Tensile	Flexural	Compressive	
Glass	6	19	52	52	18
Steel	2.0	–	12	60	20
Asbestos	9–12	18	32	73	2
Kevlar	1.9	16	44	–	17
None	0	7	13	52	0.1

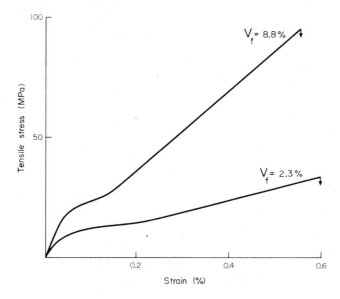

FIG. 11.17. Stress–strain curves for continuous aligned steel-cement. (After Aveston, J., Mercer, R. A. and Sillwood, J. M. (1974) NPL Conf., 93.)

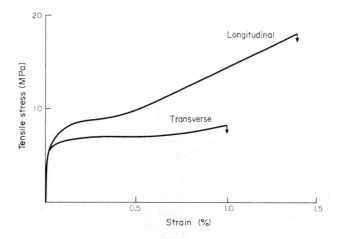

FIG. 11.18. Stress–strain curves for random glass-cement. (After Proctor, B. A. (1978) *Composites* **9**, 44.)

these composites (> 0.01) is particularly noteworthy. The difference in behaviour between longitudinal and transverse directions indicates that the glass is not completely randomly oriented. The same effects are observed with asbestos cements.

The modulus and strength of reinforced and unreinforced cements depend a great deal on the curing conditions. Air-stored reinforced cements are usually stronger than water-stored ones. Water/cement ratio is also very important, and affects the modulus of the cement more than the presence of the fibres.

Reinforced mortars and concretes have not been so exhaustively examined as reinforced cements. (Mortar is cement with sand; concrete is cement with sand and stones.) The addition of glass or steel fibres can be beneficial. Thus the flexural strength of concrete containing 1 % of 40-mm-long steel wires, 0.4 mm in diameter is 30 %–70 % better than the unreinforced concrete. The amount of the increase depends on the aggregate used. The shear strength is also increased, but for compressive strength there are conflicting reports, both increases and decreases being observed. Increases in modulus of up to 3 % have been noted for steel-concretes.

With glass-concretes, the flexural strength increases linearly with V_f up to about $V_f = 0.0075$, at a rate of about 20 % per 1 % of fibre. The rate depends on fibre length, and similar results are obtained for tensile and compressive strengths. Properties usually fall off when V_f exceeds 0.0075.

Polypropylene (fibrillated) is being used in increasing quantities to reinforce concrete. Although the fibres are not very strong, additions of up to $V_f = 0.006$ increases the strength of the concrete. Above 0.006, the effect declines rapidly, and at $V_f = 0.01$ the polypropylene-concrete is weaker than unreinforced concrete.

Glass-gypsum plaster with $V_f = 0.06$ has nearly the same strength as glass-cement, i.e. tensile 18 MPa, flexural 35 MPa, and compressive 40 MPa. The compressive strength falls linearly with glass content from about 50 MPa at $V_f = 0$, while tensile and flexural strengths increase linearly with V_f up to about 0.04.

11.2.5. Toughness

The toughness of glass and steel cement is about an order of magnitude greater than that of asbestos cement, and two orders of magnitude greater than unreinforced cement, Table 11.4. Polymers (especially fibrillated polypropylene, on account of its cheapness) are being used increasingly for toughening concrete, and carbon fibres have been shown to toughen cements.

The toughness of random glass-cement increases monotonically with increasing fibre length in the range examined (10–14 mm). With steel-concrete longer fibres have been tested, and shown to give a maximum toughness at a length of about 120 mm. Toughness increases with increasing V_f, up to about $V_f = 0.06$ for steel and glass, as shown in Fig. 11.19. With asbestos the toughness increases up to about $V_f = 0.12$.

The toughness of glass-gypsum increases linearly with V_f up to about 0.06 at which point the work of fracture is about 50 kJ m^{-2}. (In the absence of fibres it is about 2 kJ m^{-2}.)

11.2.6. Fatigue and creep

The fatigue of glass and Kevlar cements show no apparent endurance limit for tests up to about 10^7 cycles. Figure 11.20 shows the results obtained for glass in flexure and tension. Both decline by about the same proportion in the range 10^4 to 10^7 cycles (i.e. about 50 % loss). Kevlar, tested in flexure, showed about the same loss, and a marked drop in modulus was noted when the maximum stress in fatigue exceeded the cracking-stress. The fall is quite rapid initially, but slower later, the modulus eventually being reduced to one-twentieth of its initial value.

Glass does not prevent the creep of cement. Figure 11.21 shows that creep in air is roughly the same as in water for glass-cement. Unreinforced cement gives an almost identical line when plotted in this way.

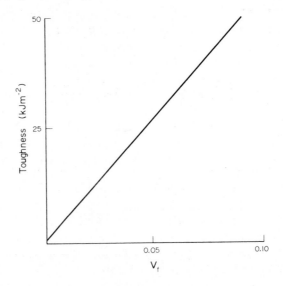

FIG. 11.19. Toughness of glass-gypsum.
(After Ali, M. A. and Grimer, F. J. (1969) *J. Mater. Sci.* **4**, 389.)

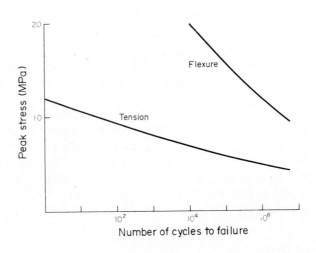

FIG. 11.20. Fatigue curves for random glass-cement.
(After Hibberd, A. P.,and Grimmer, F. J. (1975) *J. Mater. Sci.* **10**, 2124 and BRE 38/76.)

11.2.7. Environment Effects

Reinforced cements are being used in the construction of roads and buildings where the environmental agent of most importance is water, in liquid and vapour form. The resistance of plasters to water is very poor, and reinforced plasters should not be used in conditions of high humidity. However, glass-gypsum has good fire resistance.

Asbestos-cement stands up to water very well, and is used for roofing and pipes carrying water. Over very long periods (150 years) some embrittlement is observed, however, and

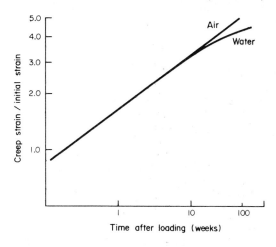

FIG. 11.21. Flexural creep of random glass-cement. (After BRE 38/76.)

chemical changes have been noted at the fibre–matrix interface, but these changes do not weaken the fibres significantly.

Glass-cement is badly affected by water. Figure 11.22 compares the fracture surfaces after 90 days in air and after 5 years' storage in water. The fibre pull-out which gave the initial high toughness and improved strength has almost entirely disappeared. This effect is not due to chemical attack and direct weakening of the fibres. Instead, the moisture causes progressive growth of calcium alumino silicate into the interstices in the fibre bundles. This increases the interfacial shear stress, and hence decreases toughness, since the fibres break rather than pull out. (Figure 7.4 shows that pull-out work can be much larger than that due to fibre failure in the crack plane.) Another factor which is important is that the fibres pull out as bundles in new composites, whereas they behave as individuals after the growth of the silicate into the interstices. This can result in as much as a thirty-fold decrease in effective "fibre" diameter, and is accompanied by a proportionate reduction in toughness (see equation 7.9). Table 11.5 shows the change in properties estimated for 20 years under various conditions. The loss in tensile and flexural strength is

TABLE 11.5. *Approximate Estimated Changes in Properties for a Glass-Cement after 20 years under various conditions.*

	Initial value	Change in property (%)		
		Dry air	Under water	U.K. weather
Flexural strength	42 MPa	−30	−45	−57
Flexural yield[†]	12 MPa	−25	+42	+25
Tensile strength	16 MPa	−12	−37	−56
Tensile yield[†]	8 MPa	0	+25	−12
Tensile modulus	25 GPa	−8	+24	+16
Toughness (Izod)	18 kJm^{-2}	−6	−67	−78

[†] End of linear region of force–distance curve.
(Results taken from Proctor, B. A. (1978) *Composites* **9**, 44–8.)

F<small>IG</small>. 11.22. Fracture surfaces of glass cement, (*a*) after 90 days in dry air, (*b*) after 5 years in water at 20 C. (Crown copyright. Reproduced by permission of the controller, HMSO.)

a reflection of the loss in toughness, since these are flaw-controlled modes of failure with this type of material. Because of this problem glass cement is not presently recommended for load-bearing applications, except for situations where the load has to be sustained over a short period, such as for permanent shuttering for reinforced concrete. (The strength is only needed while the concrete is setting.) Figure 11.23 graphically depicts the loss in strength and toughness.

The initial properties of glass-cements are also better if they are kept dry. This is in marked contrast to Kevlar-cement, which is better if kept wet. This composite withstands the effect of long-term storage under damp conditions much better than glass-cement.

Steel-cement does not suffer from this humidity problem to the same extent as glass. However, the steel must not be permitted to corrode. The cement provides an alkaline environment which inhibits corrosion, and the material can withstand very fine cracks. However stresses should be kept well below the knee in the stress–strain curve so that cracks are not formed which can permit the ingress of water. Water, especially if

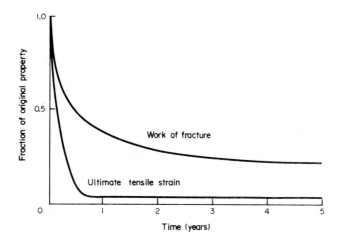

FIG. 11.23. Loss in properties on immersion in water at 20C of random glass cement.
(After BRE 38/76.)

accompanied by the chloride ions present when salt is used for ice and snow removal, can cause rapid corrosion of the steel. Loss of strength occurs in the composite only when the steel is rusted almost right through.

Further Reading

DONALD, I. W. and McMILLAN, P. W. (1976) *J. Mater. Sci*, **11**, 949.
RILEM Symposium on Fibre Reinforced Cement and Concrete (RILEM International Union Testing and Research Labs. for Material and Structures, 1975).
KINGERY, W. D., BOWEN, H. K. and UHLMANN, D. R. (1975) *Introduction to Ceramics* (John Wiley, New York).

12

Applications

THE USE of fibre reinforced materials is expanding rapidly. These materials are very versatile due to the wide choice of possible polymer matrix materials, and the variety of fibres available. Thus reinforced plastics can be designed for a specific application, by suitable choice of components, and volume fractions. In addition, many fibre-polymer combinations have the advantages found with polymers; i.e. easy mouldability, low production costs, high accuracy of moulded parts, and low moulding pressures and temperatures.

Fibre reinforced metals have not been so successful. Difficulties in manufacture, leading to excessively high production costs, mean that they can only be used for specialized applications in which material costs are of secondary importance. They are much more brittle than the unreinforced metal (unless ductile fibres are used), and their use at high temperatures is limited by chemical reactions that degrade the fibres and lead to serve loss in properties. Thermal cycling also causes degradation. In addition potential competition has spurred metallurgists to improve light metals, such as titanium, so as to keep ahead of fibre reinforced rivals.

Reinforced cements and plasters are not at the moment recommended for load-bearing applications because of the gradual loss of strength and toughness in damp environments. Fibre reinforced ceramics suffer from the same problems as reinforced metals, and are not yet successful commercially.

In this chapter particular attention will be given to applications of composites which make use of their unique advantages. In such a rapidly expanding field as this, the treatment will necessarily be far from complete, but should give an indication of the tremendous potential these materials have.

12.1. Airframes

The first use of glass-polyester was in aircraft radar domes in 1941, and in 1944 a four-passenger plane made from fibreglass was flight tested. Fibreglass is now widely used in light aircraft, and for high performance sailplanes. The low modulus of fibreglass makes it unsuitable for military and large commercial aircraft, and instead advanced composites with laminates of boron-epoxy and carbon-epoxy are being introduced. The excellent specific moduli and strength ideally suit them for the construction of high aspect ratio, thin wings, with reduced drag as well as lighter weight. Further reduction in drag results from the use of glued joints instead of riveting.

The high damping capacity of advanced composites is also an advantage, reducing noise and troublesome vibrations in control surfaces. Also manufacture can be cheaper; prepeg tape construction involves lower labour costs than riveted aluminium.

More interest is being shown in Kevlar-polymers, since these are much tougher than boron or carbon polymers. There is also some use of boron–aluminium in military aircraft. Figure 12.1 shows some B–Al leading edges for wings.

In commercial aircraft, load-carrying capacity, and hence profitability, can be increased by the use of very light structural materials. Thus the Boeing 747 has carbon-reinforced floor panels, and helicopters are being designed with Kevlar-polymer bodies.

FIG. 12.1. Boron–aluminium leading edges about 15 cm long.
(Courtesy of D.W.A. Composite Specialities.)

In space vehicles the ablation surfaces which were made from reinforced polymers are being replaced by ceramic tiles and carbon–carbon composites, which should be re-usable. Reinforced polymers are being used for many internal parts of the space shuttle, and doors and hatches.

Helicopter blades, and the rotor head assembly, are now made with reinforced plastics. The life of the blades is greatly increased, and the performance is improved as compared with metal blades. Carbon and glass fibres are used, and excellent impact resistance can be achieved, together with immunity to corrosion, and reduction in vibration and noise. The rotor heads use glass-epoxy. This makes possible great simplification in design, with only one quarter of the parts used in traditional designs. Figure 12.2 shows the composite rotor head and blades of a Dauphin helicopter, and a section of the hybrid blade of a Westland helicopter.

FIG. 12.2. Reinforced polymer helicopters driving-parts: (*a*) Section of rotor blade moulding. (Courtesy of Westland Helicopters Ltd.) (*b*) Rotor head. (Courtesy of Societe Nationale Industrielle .\erospatiale.)

Composites are also being introduced into aircraft engines. These will be discussed in the next section.

12.2. Energy Conversion and Storage

Figure 12.3 shows the profile of a typical aircraft turbine engine. At the cold end, carbon fibre reinforced polymers have been tried, with polyimide matrices being used for the warmer regions. In the centre part of the turbine boron–titanium has been used. Both should give considerable reductions in weight. However, the carbon-polymer is not yet tough enough to withstand the ingestion of medium-size birds (these can seriously endanger jet aircraft on take-off and landing) and not sufficiently resistant to erosion from rain and dirt.

At the hot end of the turbine there has been no success so far with composites, but for ground-based turbines for electricity generation, a tungsten–nickel superalloy blade is being used. The tungsten fibres are coated with tantalum carbide, and the blades are made by hot isostatic pressing. They are expected to lead to a 25 C increase in operating temperature, thereby increasing efficiency by 1 %. This area offers great potential for composites, and significant advances should be seen here in the next decade; electricity-generating turbines are not subjected to frequent thermal cycles.

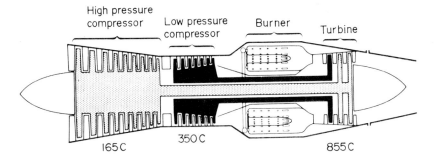

High pressure
compressor   Low pressure   Burner
                  compressor   Turbine

350 C

165 C   855 C

FIG. 12.3. Section of aircraft jet engine showing temperature regime.
(Courtesy of Pratt & Whitney.)

Nuclear reactors have severe materials problems, and silica–aluminium and carbon–carbon have been considered for the pressure tubes for fission reactors. However, neither composite has been used in a reactor. Design work is proceeding on a fusion reactor, based on the Tokamak confined gas plasma approach. Reinforced metals and ceramics will be needed because of the severe stress–temperature regime. This is another area of great potential for composites.

The large collectors used to heat water with solar energy are now being made with glass polymers (Fig. 12.4). The curved glass-polymer sheets combine high strength with the ability to transmit the appropriate wavelength of radiation to the blackened metal pipe absorbers.

FIG. 12.4. Glass-polymer solar panels. (Courtesy of Solartech Ltd.)

Reinforced polymers have been considered for energy storage in the form of flywheels. Since centrifugal forces depend on weight, the efficiency of a flywheel is determined by its tensile strength only (the extra energy associated with heavy weight is cancelled out by the extra strength required). The theoretical maximum efficiency on a volume basis is equal to the ultimate tensile strength. The best that can be achieved at the moment is a steel-polymer with a strength of 3.2 GPa. This gives a theoretical specific energy of 3.2 GJ m^{-3}. It should be possible eventually to make use of the strength of SiC whiskers, which should be capable of achieving a specific energy of 16 GJ m^{-3}, neglecting all associated equipment. This is better than the sodium–sulphur battery with a theoretical value of 3.73 GJ m^{-3}, which is being considered for electric cars. However, hydrocarbons will always be strong competitors, since octane represents an energy storage 33.7 GJ m^{-3}, and only needs a thin shell to hold it.

12.3. Ground Transport

Fibreglass is widely used for the bodywork of buses, lorries, and mobile homes. Its main advantages are good corrosion resistance, light weight, and better damping characteristics than metals. For the interior of public transport vehicles fibreglass is much used because, in addition, it is almost vandal proof.

Fibreglass is also much used for the bodies of speciality and sports cars, but the extra cost of it has prevented it being used for normal production models. However, many interior parts are made from reinforced plastics where easy mouldability and light weight are advantages. Mica reinforcement is used to an increasing extent, but for stiffening rather than strengthening.

Carbon-polymers are used in racing cars, usually in hybrids with glass. 4 kg of carbon gives great improvement in fatigue life, reduces noise, and cuts the weight down by 27 kg.

The Ford Motor Company is investigating the use of carbon in production model cars, in order to meet the U.S. Government weight restrictions for the 1980s. The body structure and frame (chassis) can be replaced, together with many small items such as drive shaft, wheels, and rear suspension trailing arm (Fig. 12.5). The weight of these components can be reduced to less than half the present weight. Doors and bumpers can be designed to withstand impact better than steel, despite the inherent brittleness of the carbon-polymer. Sandwich construction and hybridization with glass fibres are used, and the weight reduction is over 60 %. In addition, the damping effect of composites reduces the noise made by the car.

Of particular interest is the carbon-epoxy leaf spring. This is so light that a spring for a lorry can now be replaced quite easily, since it weighs only 13.6 kg instead of the 60 kg for a steel one. Its shear failure mode presents fewer hazards than the complete failure of the steel.

Rapid light transit systems being designed for cities make much use of reinforced plastics, as also do the new generation of high-speed trains. Great benefits will accrue from the reduced weight. Freight containers are also being made from reinforced polymers rather than metals. Although the composite is initially more expensive than steel or aluminium, it has proved to be more durable, and hence cheaper in the long run. Glass-polyester foam sandwiches are particularly favoured for air freight, where their light weight is a great advantage.

12.4. Marine Applications

Glass reinforced polymers are well established in the boat-building industry, having been used since the late 1940s. Their characteristics of light weight and high strength,

FIG. 12.5. Ford Ltd. body and parts made from carbon polymer. Note light weight of composite door.
(Courtesy of Ford Motor Company.)

design flexibility, and low thermal conductivity are very advantageous in this application. The monolithic seamless construction minimizes assembly problems and leakage, while maintenanace and repair costs are reduced. Their most important advantage is their excellent resistance to the marine environment. However, they generally cost more, and the combustibility of the polymer matrix can be a problem, unless the polymer is suitably treated.

Polyester resins are most commonly used for the matrix, being cheaper and easier to handle than epoxy resins. Epoxies are used only where their better resistance to weathering and moisture is required. With glass-polyester laminates the loss in strength due to water in moderate climates is some 10–15 %, increasing to a maximum of 20 % in warm climates. The strength loss occurs during the first two months, and is due to the water penetrating the resin and acting as a plasticizer. If the fibres do not have the appropriate sizing the water is absorbed on the glass, greatly weakening the fibre–matrix interface and causing a large reduction in strength. Improperly finished or under-cured resins, or laminates with excessive voids, are also more susceptible to water degradation. Thus care is needed when fibreglass is used for construction.

The glass used is usually E-glass, and the form of reinforcement used is random mats, spray-up, or woven cloths, according to the type of boat. The smallest boats are also made by moulding short glass fibre reinforced thermoplastics. Figure 12.6 shows a canoe made in this way; several years of hard use in the Canadian north have proved it to be leak proof and almost dent proof.

A very large fraction of small boats are now made from fibreglass, and the use of the material is being extended to ever larger boats. A notable example is a 47-m minesweeper.

Fig. 12.6. Glass-polyethylene thermoformed canoe. (Courtesy of Fiberglas Canada Ltd.)

Although costing more than steel, its advantages (being non-magnetic and more corrosion resistant) were worth the extra expense. For boat sizes larger than about 70 m the use of fibreglass is presently uneconomic, however.

For large boats, ferro cement is being used increasingly. Although its specific properties are much lower than fibreglass it is extremely cheap and easy to mould.

Kevlar and carbon are having a big effect on sports boat construction. The light weight and extra stiffness are the main advantages here. Kevlar is mainly used in hybrids with glass or carbon because it is otherwise prone to failure by buckling, due to its weakness in compression.

12.5. Pipelines and Chemical Plant

Fibre-reinforced plastics are ideally suited to many situations where fluids have to be handled. They are very corrosion resistant, and may be used for gaseous bromine, chlorine, and carbon monoxide, and concentrated or dilute acids and alkalis. They can match the properties of Hastlelloy C, and perform better than stainless steels in many cases. They are, however, limited by relatively low operating temperature, and their susceptibility to attack by some organic liquids.

Being lighter than metals, composite structures are much easier to transport to the site, and install. They usually require less maintenance than metals, since they do not have to be checked so frequently for corrosive damage, and in some instances their low thermal conductivity reduces the amount of insulation required.

Pressure vessels and tubes made from reinforced polymers must be lined, otherwise they start to leak at stresses which are very much less than the ultimate tensile strength of the composite. This liner can be metal (suitable for the handling of organic fluids) or polymer (generally used for inorganic fluids). They are usually made by filament winding, and the process has been well established commercially for more than a decade.

Pipes are also made from reinforced cements. Asbestos is still widely used as a reinforcement, but is being supplanted by glass, which has better flexural strength. A glass-cement sewer pipe is shown in Fig. 12.7. Reinforced cements are preferred for non-corrosive fluids, where high performance is not required, and their relatively low cost is an advantage.

Glass-polymer pipe is replacing metal in applications where corrosion occurs, for example, to carry industrial effluents containing dilute acids. In such situations glass-polyester can out-perform lead-lined steel. Epoxy resin is used instead of polyester for higher temperatures (150 C maximum, compared with 125 C for polyester) and situations where its better impact properties are an advantage.

The smoothness of the interior of glass-polymer pipes reduces the resistance to liquid flow, with the result that lines using gravity flow can have a lower slope or smaller diameter than the equivalent metal or cement lines. In addition installation costs are less because of the smaller weight.

In chemical plants it is often necessary that the materials be as fireproof as possible. Glass fibre reinforced plastics are more fire-resistant than the plastic matrix on its own because the glass will not burn. If greater fire retardancy is required than that contributed by the glass, fire retardants such as halogenated polymers or antimony trioxide are added to the resin.

Carbon fibres, because of their high cost, are only used in chemical plants where their excellent specific properties are worth the extra cost, for example in industrial centrifuges. Their electrical conductivity may also be used in applications where a vessel requires electric heat; the heating can be produced by passing an electric current through the fibres.

Storage vessels for corrosive materials are now made from reinforced polyester relatively cheaply. They have good mechanical properties and excellent durability. Chimney liners and venting ducts are also commonly made of glass reinforced polyester. Filament wound glass-polyester has been used successfully in stacks up to 400 m high, and has proved to be much more corrosion resistant than conventional liners.

The light weight of glass-polymer foam sandwich construction has been put to good use for floating roofs for oil storage, and demineralized water tanks. Installation is simplified, and cleaning can be accomplished without hindrance, since the floating roofs are light enough to be suspended from above. In addition the roof is naturally buoyant, so will not sink if punctured, and is strong enough to permit maintenance personnel to walk on it.

Concern about pollution has increased the need to concentrate wastes rather than dispersing them. Consequently, treatment plants have to handle more corrosive mixtures, and fibreglass is eminently suitable. Scrubber-demisters to handle flue gases coming from waste incinerators have been constructed from fibreglass with less difficulty and at lower cost than from stainless steel. The gases involved included hydrogen chloride and water vapour at 80 C, which few materials can withstand. A chlorine-caustic neutralizing tower 14 m high and 2 m in diameter has also been made from fibreglass. The environment included chlorine gas, hydrochloric acid, sodium hydroxide, and sodium hypochlorite, which could only be withstood by such metals as Hastelloy or titanium, which cost about eight times as much as fibreglass.

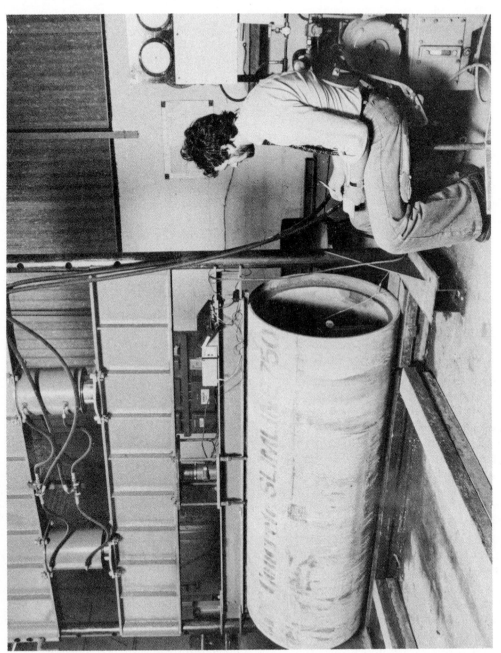

FIG. 12.7. Glass cement sewer pipe. (Courtesy of Fibreglass UK Ltd.)

Reinforced thermoplastics are also replacing metals in chemical plant. Polypropylene is one of the cheapest thermoplastics and has good corrosion resistance. Reinforced polypropylene can be injection moulded, compression moulded, extruded, etc. It has been used in a wide range of filtration applications, such as rotary drum filters, and filter press plates. Though costing more than steel, it can handle a wide range of corrosive fluids, at temperatures above 100 C. It is also used in pumps, and systems to handle corrosive electroplating solutions.

Engineers have been relatively slow to apply reinforced plastics in the chemical industry owing to their much greater familiarity with the properties of metals. However, with increasing experience with reinforced plastics, use of these materials is growing rapidly.

12.6. Buildings

A "fibre" reinforced composite has been in use in the building industry for more than a century. This is reinforced concrete, which uses steel bars with diameters of the order of 10 mm. The bars are deeply indented, normal to their axes, to provide mechanical interlocking with the concrete matrix. Their coefficient of expansion is about the same as that of the concrete matrix, so that temperature cycling does not cause problems. This material, invented in France in about 1850, is clearly a forerunner of modern composite materials which employ many of the same principles, but use reinforcement of much smaller diameter. Reinforced concrete is very widely used nowadays in the construction industry. In special cases, where extra strength is required, the steel is pre-stressed.

Asbestos cement has also been used for a long time for roofing, permanent shuttering for reinforced concrete, and for piles. It tends to be somewhat brittle, and supplies of high-quality asbestos are being depleted. Thus there is a strong incentive to replace this material.

The construction of a building using traditional methods is time-consuming and labour-intensive. With increasing labour costs, the advantages of prefabricated buildings have become more noticeable, and it is in this area that composites are having their largest effect.

Reinforced polymers are used both for the structure of the building and internal components, such as bath tubs, etc. Their use markedly decreases the weight of the structure, thus reducing the foundation size. They are much stronger and less inclined to crack. Their major disadvantage is their relatively poor fire resistance. This requires the use of modified polymers and fire-retarding fillers. They are also used for moulds for large concrete structures.

Large glass fibre reinforced polymer panels were used to construct the Bell System Building for the New York World Fair in 1963. They were about 12 m long and 4 m wide, and were backed with steel reinforcement and asbestos cement insulation. They could be made quickly, and with consistent properties and good finish, in the factory, and assembly at the site was expeditious. Figure 12.8 shows a glass sandwich structure for walls, floor, and roof.

More recently sandwich panels have been made which provide all the strength needed for small buildings without the use of steel. For example, a glass reinforced polyester skin, honeycomb core, material has been used for the sides of an A-frame house. The fibreglass provided the strength needed, while the core gave good insulation. The floor panels were

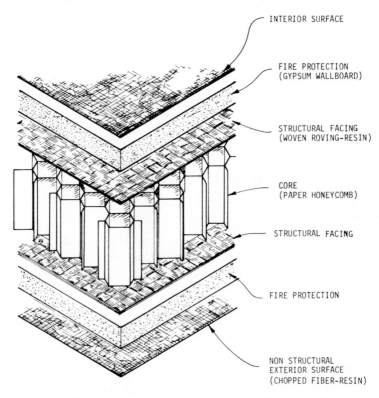

INTERIOR SURFACE

FIRE PROTECTION
(GYPSUM WALLBOARD)

STRUCTURAL FACING
(WOVEN ROVING-RESIN)

CORE
(PAPER HONEYCOMB)

STRUCTURAL FACING

FIRE PROTECTION

NON STRUCTURAL
EXTERIOR SURFACE
(CHOPPED FIBER-RESIN)

FIG. 12.8. Glass-polymer sandwich structure for walls, floor, and roof. (Dietz, A. G. H. (1974) *Composite Materials* **3,** 290 (Ed. Noton, B. R.) Courtesy of Academic Press.)

also prefabricated from fibreglass, and the electric wiring and rough plumbing were installed in the floor systems at the factory. To complete the house up to the stage of finishing the shell and interior walls and floors, and the plumbing and wiring, only took 150 man-hours.

The speed of assembly that can be achieved with reinforced polymers could very greatly assist rehousing after a disaster. A small fibreglass house-shelter has been designed which can be constructed in a few hours by first spraying the fibreglass onto a levelled piece of ground for the floor, and then, for the walls and roof, spraying the material onto an inflated silicone bag placed thereon, to which are attached window and door frames. Removal of the silicone bag, and hanging the windows and door, completes the shell of the building. The design provides about $53 \, \text{m}^2$, at a cost of only about $\$70/\text{m}^2$ (in 1978) including fixtures and furnishings.

Buildings supported by air pressure also use glass-polymers. Air supported structures can be very large, e.g. $15,000 \, \text{m}^2$, and are anchored only at the periphery. They need only very small air pressures to support them, i.e. about $0.2 \, \text{kPa}$ ($0.002 \, \text{atm.}$). Figure 12.9 shows a hockey rink under a glass polyvinyl chloride skin.

FIG. 12.9. Glass-polymer inflated hockey rink. (Dietz, A. G. H., (1974) *Composite Materials* **3**, 290 (Ed. Noton, B. R.) Courtesy of Academic Press.)

Glass-polymers are finding increasing employment replacing wood and steel for the shuttering used to hold concrete in shape while it is setting. The glass-polymer is sufficiently stiff and strong to support the wet concrete, and light enough to be easily manhandled. Extra reinforcement is being provided by carbon fibres, strategically placed.

Glass-cements have mainly been used in non-load-bearing situations, for example as cladding panels for building exteriors and as attractive earth-retaining wall panels. They weigh 25 % less than comparable concrete panels, can be fixed by screws and cut with a saw.

The high initial strength of glass-cement made possible the construction of the shell structure shown in Fig. 12.10. This 31-m diameter structure, built for the 1977 Spring Flower Show in Stuttgart, W. Germany, is part of an experimental study in the use of these materials.

FIG. 12.10. Glass-cement shell structure spanning 31 m with 10 mm thickness.
(Courtesy of Fibreglass UK Ltd.)

Glass-gypsum plaster is used for inner walls, for plastering or decorative purposes, where something tougher is required than the normal unreinforced plaster, e.g. in school and public buildings. This material has also been used to make prototype light-weight floor panels for internal use.

Glass and steel concretes are being used mainly on an experimental basis, for road surfaces on bridges and areas of heavy use, and also for multi-storey car parks. Their cracking resistance, and lighter weight for the same strength, are the principal reasons for these applications.

12.7. Medical Applications

The economics of the medical use of materials is quite different from that of most other applications. The major costs in medicine are the facilities: hospital, equipment, etc., and the labour: surgeons, anaesthetists, and other highly paid specialists, together with nurses and support staff. The amount of material in a medical application is usually quite small, less than 1 kg, so the cost of even the most expensive materials is usually insignificant compared with the other costs. In addition, there is the time and suffering of the individual requiring the device, and expensive composites can often assist speedy recovery.

A new material cannot be accepted immediately for medical use. A device designed for internal (prosthetic) use has to be checked for unfavourable reactions, both short-term and long-term. The same is true for external (orthotic) devices, though the requirements are generally much less stringent.

Fibreglass is rapidly supplanting wood, leather, and steel for braces (or callipers) for arms and legs. The braces are stronger, lighter, more comfortable, and less noticeable. No straps are needed, and improved cuffs, having air holes, and padded with washable foam, can easily be produced.

Fibreglass is also used for artificial legs and arms. It may soon be supplanted by carbon-polymer because this has better specific properties. An important advantage for reinforced polymers is their easy mouldability, in view of the wide range of shapes and sizes required. The prostheses are made from plaster moulds of the parts, using woven reinforcement with open, easily shaped weaves. These prostheses can very easily be made to mimic the appearance of human limbs.

A glass-thermoset is being tested as a replacement for traditional plaster casts. It makes a lightweight, strong, and ventilatable cast, that is not harmed by water. In its present form, however, the material irritates the skin, and setting is slow. An alternative may be a polymer fibre-thermoplastic that can be shaped after dipping in hot water. Its setting time of 5 s is, however, rather short for the necessary shaping.

Internally, it is hoped to replace Vitallium (a cobalt alloy with 20 % chromium and a few per cent of molybdenum) which is used for bone splits, hip and knee joints, etc., with carbon-polymers. Figure 12.11 shows a prototype carbon-epoxy splint. A great

FIG. 12.11. Carbon-epoxy splint, $13\frac{1}{2}$ cm long. (Courtesy of Bradley, J., Stoke Polytechnic.)

advantage of the composite is the possibility of matching the modulus of the bone material. This prevents the de-stressing, and consequent dissolution of the bone adjacent to the joint, that follows the use of the high modulus metal. Much further work is needed here to solve problems of inflammation and necrosis of surrounding tissue due to the presence of the polymer.

Recent observations suggest that plaited continuous carbon fibre can be used to make new tendons. The body eventually replaces the carbon with collagen to produce a natural tendon.

Carbon-polymethyl methacrylate is being used for a bone cement. It has superior fracture resistance and greater durability under load than the unreinforced polymer. Animal tests have shown its usefulness in fixing internal splints and other prostheses, and its use is being extended to human subjects.

The combination of excellent mechanical properties and the very low X-ray absorbtion characteristics of carbon fibres is providing a unique opportunity for the development of better X-ray analysis and treatment devices. The angiographic technique (which involves the injection of a radio opaque fluid into the bloodstream, for the location of growths and foreign bodies) needs X-ray pictures to be taken in quick succession (up to six per second). The film has to be placed and moved on very quickly, but held precisely. The key to doing this has been the development of carbon-epoxy compression plates. These can be very precisely moulded from prepreg tape, and the plate combines excellent fatigue properties with low X-ray absorbtion and good elastic properties.

The computed tomogamy X-ray body scanning technique is used to provide a three-dimensional view of the body. This requires a support which provides great rigidity, so that a person lying on it does not deflect it significantly, and at the same time it must have low and consistent X-ray absorbtion for radiation passing in any direction in a plane. A carbon-epoxy foam sandwich structure has been developed for this which is highly successful, allowing X-ray resolutions that were not previously attainable. A less exacting requirement is for thereapeutic X-ray table tops. These also are made with carbon-epoxy foam sandwiches, and one is shown in Fig. 12.12.

In these X-ray applications, no other material has been found which comes anywhere near the excellence of carbon-polymers.

12.8. Sports Equipment

Materials cost is often a secondary consideration in sports equipment. Thus carbon and boron fibres are used in golf clubs and tennis and badminton racquets. Fibreglass is also widely used. The fibreglass fishing rod has almost entirely superseded the wooden one. Skis often contain fibreglass, and so do ski poles, bows for archery, gun stocks and butts, hockey sticks, and helmets and face masks. These applications all represent advances over traditional materials due to better durability and improved specific mechanical properties.

12.9. Other Uses

Fibreglass and other reinforced polymers seem destined eventually to take over nearly all uses of metal which do not require the unique properties of the more useful metals, i.e.

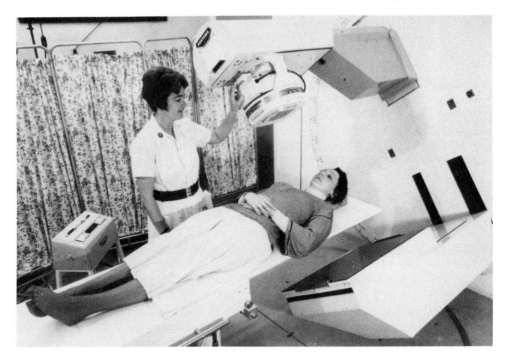

FIG. 12.12. Carbon-epoxy support for X-ray therapy, Lyons, B. R. and Molyneux, M. (1978)
Proc. ICCM2 1474, Courtesy of the Metallurgical Soc., AIME.)

ductility, hardness, high-temperature resistance, and conductivity. Unreinforced plastics
have already supplanted metals in many applications where the low modulus and strength
of the polymer are not a disadvantage. (A good reason for the use of polymers in our
increasingly energy-conscious era is the low energy required to produce and process them,
as compared with ferrous alloys and aluminium, the two most commonly used metals.)
Now that reinforced polymers of good quality can be produced reliably and cheaply,
these can be expected to take over in areas where moderate to high modulus and strength
are required. In addition, the unique properties of some composites (for example the lack
of thermal expansion of carbon reinforced polyesters and epoxies) can sometimes increase
their advantages over metals at moderate temperatures.

Glass fibre reinforced plastics are widely used in agriculture, home appliances, business
machines, electrical and electronic hardware, and in materials handling, as well as in the
areas already described in more detail above.

Carbon fibre reinforced plastics, being a much more recent development, are much
less widely used at present. A use which takes advantage of the negligible thermal
expansion of this material is a radar reflector dish used as part of an antenna. This is a
sandwich structure, with an aluminium honeycomb core. The thermal stability of the
carbon-polymer ensures that the antenna remains accurately tuned over a range of
ambient temperatures.

Another use of carbon fibres that takes advantage of its unique properties, this time
light weight, is for moving parts in textile machinery. New machine designs using carbon
pultruded bars (which can be made with greater precision than metal constructions), have

increased weaving speeds by a factor of ten to twenty compared with conventional machines. The machines also contain injection moulded carbon fibre reinforced nylon. These mouldings can be made very accurately, allowing very precise alignments of the parts during machine operation. In another textile-processing machine an injection moulded nylon traverse guide is used. It is shown in Fig. 12.13. The speed of operation of this machine is limited by friction and wear of this part, and the good friction and wear characteristics of the carbon–nylon are a great advantage. In addition, the carbon conducts the frictional heat away from the sliding face more efficiently than the unreinforced polyamide used previously. These advantages are also made use of in carbon polytetrafluoroethylene (PTFE) bearings used in other machines.

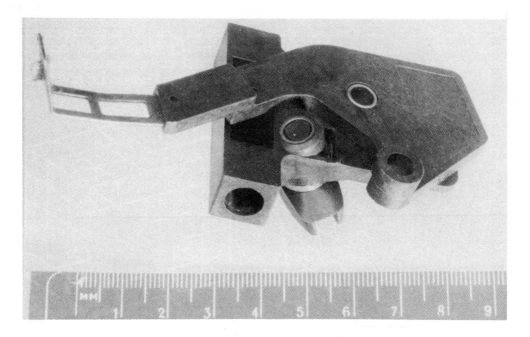

Fig. 12.13. Carbon–nylon traverse guide for textile processing. (Trewin, E. M. 1978) Proc. ICCM2 1474, Courtesy the Metallurgical Soc., AIME.)

Brake and clutch linings for cars, trucks, buses, and trains, shown in Fig. 12.14, are mostly made of asbestos reinforced thermosets. The asbestos provides strength. The polymer matrix behaves as a visco-elastic fluid, and it is this that controls the friction of the brake lining, although the asbestos has a high friction coefficient. A wide range of fillers is used to modify the friction and wear characteristics of the lining to suit the end use. In addition, different poylmers are used. It is possible to use this type of brake up to temperatures of about 300 C. This is because the time at high temperature is usually short. The cermets used for very heavy duty applications which involve temperatures up to 400 C, such as on aircraft wheels, are now being replaced in some applications by carbon–carbon. This material can operate well when red hot, and is, therefore, very effective at dissipating the heat generated by the large amount of kinetic energy absorbed.

Fibreglass is still used for radomes and some of these are very large nowadays; the

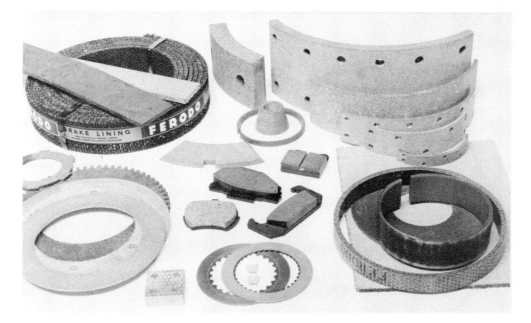

FIG. 12.14. Asbestos-polymer friction materials. (Courtesy of Ferodo Ltd.)

frontispiece shows the CN Tower in Toronto, which was built primarily as a communications tower and is the first free-standing tower to exceed 550 m in height. It houses antennae for numerous microwave links, five television channels at > 1 Mw and five F.M. channels at 40 kw each.

The radio and TV antennae are housed inside fibreglass tubes up to 7.5 m in diameter with 4 cm wall thickness and 105 m high, shown in Fig. 12.15. Apart from its excellent microwave propagation properties, the material affords excellent protection from the weather at this height. This can sometimes be very severe. It is built to withstand winds of 400 kmhr^{-1} and was undamaged by 190 kmhr^{-1} winds in February 1978 which did a great deal of damage to buildings in Toronto. Another very important property is that ice and snow do not stick to this material. There is thus no danger of it icing up and the built-up ice blocks subsequently falling more than 300 m to earth, and endangering passers-by.

Although the future for reinforced polymers is assured, reinforced metals and ceramics have not yet matured. This is almost certainly only a matter of time. Technical problems, unlike human ones, always seem to be solvable.

Further Reading

LUBIN, G. (ed.) (1969) *Handbook of Fiberglass and Advanced Plastics Composites* (Van Nostrand Reinhold, New York), Chapters 21–29.
Proceedings, ICCM1, 1976.
Proceedings, ICCM2, 1978.
NOTON, B. R. (ed.) (1974) *Composite Materials*, **3,** Applications of Composite Materials (Academic Press, New York).

FIG. 12.15. Glass-polyester radome atop Toronto's 550-m communications tower. (Courtesy of CN Tower Ltd.)

Appendix A

Symbols used in text

WHERE possible, standard nomenclature has been used. Duplication could not be avoided entirely, but each symbol is described fully in the appropriate part of the text.

A	cross-sectional area; also constant (Chapter 2); also $A \simeq 5.5\ \tau_{my}/\sigma_{fu}$ (Chapter 7)
B	constant
C	stiffness; always has subscripts indicating directions e.g. C_{16}, C_{ij}
D	constant
E	Young's modulus; usually has subscripts indicating directions or materials, e.g. E_x, E_3, E_f, E_m, E_m^*
E_L	$V_f E_f + V_m E_m$
E_r	Young's modulus for randomly aligned fibre composites
E^*	secant modulus
F	force
G	shear modulus; usually has subscripts indicating directions or materials, e.g. G_{xy}, G_{23}, G_f, G_m
$\mathbf{G}$	work of fracture; usually has subscripts indicating mechanisms
I	moment of area
K	bulk modulus
$\mathbf{K}_{1c}$	fracture toughness for the opening mode of fracture
L	length; fibre length is $2L$; matrix block length is $2L_m$
M	moment of forces
N	number of fibres crossing unit area of crack face
P	pressure
P_f	packing factor
Q	reduced stiffness; subscripts indicate directions
$\overline{Q}$	transformed reduced stiffness; subscripts indicate directions
R	radius; distance from centre
S	compliance; subscripts indicate directions
$\overline{S}$	transformed compliance; subscripts indicate directions
T	temperature
$[T]$	transformation matrix for rotation of axes
U	energy or work
V	volume fraction; subscript indicates material or form
W	dislocation width
a	adhesion coefficient
a_1	interatomic distance
a_2	distance

a_c	crack length
a_o	length of unit cell side
b	Burgers vector
d	fibre diameter or thickness
d_1	thickness
d_2	thickness
e	volume expansion
h	length or interplanar distance
k_1	constant
k_2	constant
l	length
l_c	critical length
m	Gruneisen constant (Chapter 2 only)
m	fraction of fibre subject to slip
n	Gruneisen constant (Chapter 2 only)
n	$\{2E_m/E_f(1+v_m)\ln(P_f/V_f)\}^{\frac{1}{2}}$
n^*	$n\{E_m^*/E_m\}^{\frac{1}{2}}$
p	$2\mu v_2 msE_m/E_f$
r	half the fibre diameter
r_c	crack tip radius, notch radius
s	fibre aspect ratio $= 2L/d = L/r$
s_c	critical fibre aspect ratio
s_m	aspect ratio of cracked pieces of matrix; $s_m = 2L_m/d$
$\bar{s}$	$s(1-m)$
t	thickness
u	displacement; subscript indicates direction or material
u_R	displacement at radial distance R from fibre axis
v	displacement in the y direction
w	displacement in the z direction
w_1	weight
w_2	weight
x	direction
y	direction
z	direction
α	thermal expansion coefficient
β	angle
γ	shear strain; subscripts indicate shear plane or material
δ	small change in a variable, e.g. δT
ε	strain; subscripts indicate directions or material, or limit
η	mutual influence coefficients; $\eta_{xyx} = \bar{S}_{16}E_x$, $\eta_{xyy} = \bar{S}_{26}E_y$
θ	angle
μ	friction coefficient
ν	Poisson's ratio; subscripts indicate directions or material
ρ	density
σ	tensile stress (compressive stress is $-\sigma$); subscripts indicate direction or material, or limit
τ	shear stress; subscripts indicate directions or material, or limit

ϕ angle of axis rotation
χ_1 strength factor for fibre randomness
χ_2 strength factor for fibre length
χ_p strength factor for platelets
ψ surface energy

Subscripts

0   number, usually indicating an equilibrium value
1, 2, 3, 4, 5, 6   direction or number
a adhesive; also distinguishes fibre type in a hybrid composite
b distinguishes fibre type in a hybrid composite
c centre (σ_{fc}) or with u (e.g. σ_{1cu}) used for compression; also refers to cracks (a_c, b_c, r_c),
          cracking (V_{fc}, m_c) and critical values (l_c, s_c)
e stress near end (σ_{fe}) or elastic shear (τ_e)
f fibre
i interfacial stress (τ_i, σ_i) intermediate stress (σ_{fi}, σ_{mi}); also index
j joint; also index
m matrix
max   maximum
min   minimum
p platelet (Chapter 8) or gross slip ($\varepsilon_{1p}, \sigma_{1p}$) or pull-out ($1_p$)
R property at radial distance R from fibre centre
r radial
s surface ($\tau_s, \mathbf{G}_s$) onset of slip ($\varepsilon_{1s}, \sigma_{1s}$)
t total
th   theoretical
tip   tip (crack tip stress)
u ultimate (stress)
x direction
y direction (single subscript) yield value (second subscript)
z direction
θ direction

Appendix B

SI Units

SI means Systeme Internationale d'Unites, and is an internationally agreed set of units. It is rapidly gaining world-wide acceptance, and is replacing local systems such as the British Imperial one, and its N. American cousin.

It is a modified form of the MKS system (Metre-Kilogram-Second) which was introduced in the late 1940s as an alternative to the cgs system (centimetre-gram-second) which had the disadvantage that electromagnetic and electrostatic secondary units (for example for potential difference and current) were many orders of magnitude different, and were inconsistent with the practical volts and amperes used every day. (The MKS system did not have that disadvantage, nor does SI).

In SI the basic units include metre (m), kilogram (kg), second (s), and temperature in Kelvins (K). Temperature can also be expressed in Celsius (C) but the degree sign has been dropped. To convert K to C subtract 273.2.

SI makes a clear distinction between mass and force. For force there is a secondary unit (i.e. one that can be expressed in terms of the primary units). It is called the Newton (N) and is the force required fo accelerate 1 kg at rate of 1 ms^{-2}. Its units are therefore kgms^{-2}. The earth's gravitational field exerts a force of 9.81 N on a mass of 1 kg at the earth's surface. (The force on the kilogram at the surface of the moon is about 2 N).

When writing the units, slashes are not used for inverse units. Instead a negative power is used, e.g. ms^{-1} is used for speed, rather than m/s.

Powers of ten should not be expressed explicitly. Instead a symbol is used. These are listed in Table B1. Sometimes c is used for 10^{-2}, as in cm, but this should be avoided wherever possible. Low speeds, for example, are given in mm s^{-1} (millimetres per second), higher speeds as m s^{-1} (metres per second) and very high speeds as km s^{-1} or Mm s^{-1} (kilometres per second or megametres per second).

TABLE B1. *How Orders of Magnitude are expressed in SI*

Power of ten	Name	Symbol
−18	atto	a
−15	femto	f
−12	pico	p
−9	nano	n
−6	micro	μ
−3	milli	m
3	kilo	k
6	mega	M
9	giga	G
12	tera	T

The axes of graphs should never have statements like: speed; $m s^{-1} \times 10^5$. This is confusing. Use either $km s^{-1}$ or $Mm s^{-1}$, whichever gives the fewer zeros.

In this text stresses and pressures are always given in Pascals (Pa). This is another secondary unit; $1 Pa = 1 N m^{-2}$. It is a very small unit indeed; atmospheric pressure is about 0.1 MPa. Thus we have normally to use MPa or GPa.

Expansion coefficients are given in μK^{-1} (i.e. micro metres of expansion, per metre original length, per degree Kelvin). Densities are given in the unfamiliar units of Mgm^{-3}, but since $1 Mgm^{-3} = 1 gcm^{-3}$ this should present few problems.

Table B2 lists some useful conversion factors. To make the meaning of the conversions clear, slashes separate the Imperial and SI units, and orders of magnitude are expressed as powers of 10.

TABLE B2. *Conversion Factors between SI and Imperial Units. To Make the Conversion Multiply by the Appropriate Factor*
The factors are accurate to at least one part in 1000.

	Multiplication Factor			
Quantity	SI to Imperial		Imperial to SI	
Length	3.281	ft/m	0.3048	m/ft
	39.37	in/m	0.02540	m/in
	0.03937	in/mm	25.40	mm/in
Area	10.76	ft^2/m^2	0.09290	m^2/ft^2
	1550	in^2/m^2	0.6452×10^{-3}	m^2/in^2
	1.550×10^{-3}	in^2/mm^2	645.2	mm^2/in^2
Volume	35.31	ft^3/m^3	0.02832	m^3/ft^3
	61.02×10^3	in^3/m^3	16.39×10^{-6}	m^3/in^3
	61.02×10^{-6}	in^3/mm^3	16.39×10^3	mm^3/in^3
Mass	0.9842	ton/tonne	1.016	tonne/ton
	2.205	lb/kg	0.4536	kg/lb
	35.27	oz/kg	0.02835	kg/oz
Density	62.42	$lb ft^{-3}/Mg m^{-3}$	0.01602	$Mg m^{-3}/lb ft^{-3}$
Force	0.2248	lbf/N	4.448	N/lbf
Stress and Pressure	0.1450×10^{-3}	psi/Pa	6.895×10^3	Pa/psi
Thermal expansion	0.5556	$°F^{-1}/K^{-1}$	1.800	$K^{-1}/°F^{-1}$

Problems and Answers

Chapter 1: Problems.

You are strongly recommended to solve these problems in the order given. Data needed will be found in tables in this and other chapters of the book.

1.1   Calculate the breaking strains of pure aluminium, glass, concrete, epoxy resin and polycarbonate. Assume that all these materials are perfectly elastic.

1.2   Calculate the weight of a rod of maraging steel, and the weight of Douglas fir, both 30 m long, and which, when hanging vertically, will just support a man weighing 75 kg, hanging on the lower end. Assume the density of Douglas fir is the same as that of Sitka spruce.

1.3   A bar of metal has to withstand a large temperature decrease, to a minimum temperature of 20 C, while remaining the same length. What are the corresponding maximum temperatures for bars made from aluminium alloy and tungsten? (Thermal expansion coefficients are given in Tables 3.3 and 9.6.)

1.4   A ship 165 m long is moored against a quay by a steel hawser at the bow, and a nylon rope at the stern. Both moorings are at right angles to the quayside and have a length of 11 m and a diameter of 0.127 m. They are unstressed when the ship's side is in contact with the quay. An offshore wind springs up and exerts a force of 20 MN on the ship normal to its side. What angle will the ship turn through as a result? How far will the stern section be from the quay? (Use data from Table 3.1.)

1.5   A steel girder, weighing 2.8 tonnes is supported at the centre by a vertical Kevlar rope 1.8 cm diameter, and at the ends by vertical terylene ropes 1.3 cm diameter, equal in length to the Kevlar rope. Calculate the stresses in the ropes. (Use data from Table 3.1.)

1.6   A steel rod is held at $25°$ to the vertical by a vertical 0.5 mm diameter nylon filament attached to its lower end. The upper end of the rod is held by a frictionless pivot at the same level as the upper end of the nylon filament. The rod has a diameter of 3.8 cm and a length of 1.7 m. If a heavy weight is hung on the rod, so that the end is pulled down, and the rod becomes approximately vertical, will the nylon be stretched to such an extent that its stress, calculated assuming that the material behaves elastically, exceeds its ultimate tensile strength.

1.7   A load of 2.7 tonnes was to be supported by a strip of steel, fastened to another strip of steel by an overlapping joint, held together by steel rivets, 6.35 mm diameter. However, the wrong steel was used for the rivets and in the riveting process, all but one of them broke. The strength of the steel was 190 MPa, and it was too brittle. The only material that was available was white pine, cut from a local tree. How many wooden rods of the same diameter would have to be used in place of rivets to prevent the steel rivet from breaking when the load was applied. Would this number of rods make the steel rivet redundant? Assume that the shear strengths of the steel and wood are half their respective tensile strengths, and that the stress–strain curve of the steel is linear up to the fracture stress. The shear modulus of the wood is given with sufficient accuracy by dividing the Young's modulus by 2.6. (Table 6.1 gives Poisson's ratios.)

1.8   A sapphire crystal disc 3.18 mm diameter and 1.00 mm thick is sandwiched between two glass sheets 0.178 m in diameter. The glass is also held apart by a 1.05 mm thick natural rubber ring at the periphery of the sheets, and the space between is evacuated. What would the width of the ring have to be to prevent the air pressure on the glass from breaking the sapphire, assuming the compressive strength of the sapphire is the same as high density alumina, and that the glass does not break?

1.9   A circular bar has been made from a single crystal of tin with the basal plane only $3°$ out of alignment with the plane normal to the axis of the bar. If shear can occur on the basal plane at a stress 7.0 MPa and cleavage can occur by separation at the basal plane at a stress of 60 MPa, determine the mode of failure when the bar is tested in a tensile machine. Calculate both the stress required, and the load in kg, if the diameter of the bar is 0.62 cm diameter.

It can be shown by differentiation of equation 1.8 that the maximum and minimum tensile stresses are at angles ϕ to the y axis where $\tan 2\phi = 2\tau_{xy}/(\sigma_y - \sigma_x)$. Use this result to solve the following three problems.

1.10 A house near Bangkok was supported by four unreinforced concrete stilts. It was designed so that the concrete was under compression at a stress equal to its tensile strength. What would the speed of the wind in a typhoon have to be to cause failure of the stilts? The building presented on area of 42 m² to the wind, and it weighed 35 tonnes. Neglect the moments of the wind forces, and consider the stilts as subject to shear from the wind force and compression from the building weight. You may assume that the pressure exerted by the wind is ρV^2 where V is the wind velocity and ρ is the density of the air, which has a value of 1.29 kg m⁻³.

1.11 A thin walled glass tube was used to connect a motor to a stirrer in a chemical plant. The shear stress in the glass was designed to be one quarter of its tensile failure stress. Due to a design fault the glass was also under a tensile stress along the length of the tube, and it broke. Calculate the minimum load that was needed to break the tube. The diameter of the tube was 3.0 cm, and the wall thickness 1.2 mm.

1.12 A link connecting a brake lever and the brake has a joint in it where two pieces of aluminium are held together by a 3.0 mm diameter bolt. The bolt was brittle because of having had the wrong heat treatment, and fractured at a stress of 90 MPa. The bolt was screwed up tightly so that the pieces of steel were held together with a force of 300 N. If the coefficient of friction of the aluminium surfaces is 0.3, what would be the maximum force the joint could transmit?

It can be shown, by differentiation of equation 1.10 that the maximum and minimum shear stresses are at angles ϕ to the y axis where $\tan 2\phi = (\sigma_x - \sigma_y)/2\tau_{xy}$. Use this result to solve the following three problems.

1.13 The house near Bangkok (q. 10) had its concrete supports replaced by steel ones. The steel was ductile and had a compressive failure stress equal to its tensile failure stress of 150 MPa. If the supports were designed to be under a compressive stress equal to one half of the material strength, what wind speed would be required to cause failure of the supports.

1.14 It was decided to use stainless steel with a tensile strength of 270 MPa to replace the broken glass tube in the chemical plant. (q. 11). The shear stress in the steel was designed to be one half the shear strength of the steel. What would be the minimum load needed to break this tube? (Note: you must first calculate the wall thickness of the tube; it had the same diameter as the glass one).

1.15 The bolt connecting the brake lever to the brake (q. 12) was replaced by one that was correctly heat treated, was ductile, and had a strength of 310 MPa. What would be the maximum force the joint could transmit with the new bolt?

1.16 A car is travelling along a paved country road in the spring and has to stop suddenly at an intersection. The road has been damaged badly by the winter frosts, and the asphalt is broken up into pieces which are the same size as the area of contact of the car tyre on the road. It being a cold spring morning the cracks are open near the surface, but closed further down. Determine whether the asphalt will fail beneath the car wheels, given the following data: weight of car = 1.7 tonne; coefficient of friction of tires on road = 1.3; shear strength of asphalt = 3.2 MPa; tensile strength = 0.12 MPa; thickness = 2.72 cm. At the instant of stopping tire pressure at the front is 2.2 bar, and the deceleration is sufficiently great that substantially all the car weight is supported by the front wheels. (1 bar = 100 kPa.)

1.17 The materials to bridge a small ravine, 6.0 m wide, in the mountains, have to be as light as possible in order to facilitate transport. Two horizontal, parallel, wide flange beams are to be used as the main supports, since this is the most efficient shape. They must each be 12 m long and the cross-sectional area of material in them, A, and their depth, h, must be designed so that the maximum stress does not exceed one-half of the breaking strength, and they are stiff enough that under a maximum moment of 30 tonne-metres on each support (due to the passage of a loaded vehicle) the radius of curvature of the beams has a value of 60 m. The moment of area of the beams is $I = 0.7 Ah^2$ and the maximum stress in the beam is $3M/Ah$. Develop equations for A and h, and compare the relative merits (so far as weight is concerned) of steel, aluminium alloy, and titanium alloy.

1.18 An economically inclined designer decided that a water tank could be supported by three symmetrically disposed vertical pipes, one of which could also serve for the inlet water and the other two for the outflow. He used polyvinyl chloride for tank and pipes, which had a strength of 40 MPa, and was ductile, (i.e. it had a shear failure mode). He made the pipes 10.2 cm in diameter, and made the wall thickness 1.3 mm. This, he calculated, was enough, with a 20% margin for safety, when the compressive stresses resulting from the weight of water in the tank were considered. He also checked to see if pipes with this wall thickness could withstand the tensile stresses due to the water pressure. In the inlet pipe this was 0.62 MPa. (The circumferential stress, σ, due to pressure P in a thin walled tube is $\sigma = Pr/t$ where $2r$ is the tube diameter, and t is the wall thickness). This was found to be more than adequate, so he did not check the effect of the combined compressive and tensile stresses. When tested, it was found that failure occurred before the tank had filled up. Explain why, and calculate the fraction filled at the instant of failure.

Chapter 1: Answers

(Note: answers are rounded off to the appropriate numbers of significant figures.)

1.1   0.85, 1.00, 0.070, 24, 26 millistrain

1.3   Aluminium alloy 420 C, tungsten 1040 C

1.5   Terylene 0.99 MPa, Kevlar 110 MPa

1.7   217 rods needed for steel with a modulus of 212 GPa; steel is redundant

1.9   Cleavage 60.2 MPa, 185 kg

1.11 7.4 kN

1.13 267 km h^{-1}

1.15 1.2 kN

1.17 $h = R\sigma_u/4.2\,E$; $A = 25.2\,ME/R\sigma_u^2$; figure of merit $= \sigma_u^2/E\rho$; steel/aluminium alloy/titanium alloy $= 2.4/2.2/3.6$. (M = moment, R = radius, σ_u = strength.)

Chapter 2: Problems

You are strongly recommended to solve these problems in the order given. Data needed will be found in tables in this and other chapters of the book.

2.1   Calculate the theoretical cleavage strength for nickel, which has $m = 7$, $n = 4$ and $E = 200$ GPa.

2.2   Calculate the strain in silicon at the onset of cleavage at the theoretical maximum stress, given that $m = 12$ and $n = 8$ for this material.

2.3   Copper is face-centred cubic with $E = 130$ and the same values of m and n as nickel. At what stress does the theory indicate it should fail, and what would be the expected mode of failure?

2.4   Calculate the theoretical cleavage strength of iron, following the same reasoning as in section 2.1, but using a parabolic force-distance law, instead of the sine law given in equation 2.1. For iron $E = 212$ GPa. (Note: you must first show that the parabolic law has the form $\sigma = k_3(3a_1x - x^2 - 2a_1^2)$.)

2.5   If we assume that the dislocation width can be calculated from the Peierl's stress, what would the value be for pure annealed aluminium yielding in tension at 57 MPa, with $b = 0.25$ nm?

2.6   Estimate the width of a dislocation in alumina from the hardness, if its Burgers vector is 0.475 nm.

2.7   A specimen with a notch in its centre, 3.70 cm long, was used for measuring the toughness of a magnesium alloy. The specimen was fatigued before testing, and the notch increased in length due to the formation of very sharp cracks at each end, 0.17 cm long. When tested, the crack started to propagate when the stress was 63 MPa. What was the work of fracture of the specimen?

2.8   A round tungsten bar, 2.51 cm diameter, has a very sharp crack in it which is 1.1 mm deep. If it is loaded up till it breaks, what weight would it be supporting at the moment of fracture?

2.9   If the rod described in question 2.8 was replaced by one of maraging steel with the same load carrying capacity when unnotched, what weight would it support?

2.10 An oblong steel beam, used to support a small bridge, failed by brittle fracture. It was found to have been incorrectly heat treated, and had a work of fracture of only 10.1 Jm^{-2}. Failure was initiated by a crack in the lower surface of the beam, 0.71 mm deep and 1.31 m from the centre. It was designed to bear a maximum stress of 220 MPa. Compare the weight needed to cause failure and the design maximum weight. The weight is supported by loading the beam at its centre, the beam has a span of 4.8 m, a width (b_1) of 0.051 m and a depth (d_1) of 0.21 m, see Fig. 1.12. (Hint, to calculate the stress at the cracked section, calculate the internal bending moment, M, from the moment of the applied forces, and use $MR = EI$. The strain near the surface, and hence the stress, can be calculated from R; you may assume that the region around the crack is at approximately constant stress; alternatively, the equations on p.120 may be used.)

2.11 Estimate the depth of deformed zone at each crack surface for an aluminium alloy which has fractured with $G = 260$ kJm^{-2}. The yield strength of the material is 120 MPa, and X-ray analysis showed that the plastic strain in the deformed zone was 0.98. Assume the material is perfectly plastic.

2.12 A rod 0.95 mm diameter was used to support a 5.6 kg load on a machine used to measure creep. The rod was made from brass, and was strong enough to support a load of 16.7 kg. It was, however, brittle, and failure at the maximum load was caused by small surface flaws, 7.6 μm deep. Because of this brittleness, care is needed when loading the machine, so that the 5.6 kg weight is not released at a level much higher than its working position. What is the maximum height that the load can fall through without the wire breaking? For the brass $v = 0.350$ and $E = 101$ GPa.

2.13 A ships hatchway has a tiny crack at one of its corners where it is welded to the steel deck plates. If the crack extends into the plates for a distance of 0.80 mm, and is oriented at right angles to the stress in the deck, what would the stress have to be if the work of fracture of the steel is 2.65 kJm^{-2}? The hatch is 4.00 m square and has corners rounded to 25 mm, and has one side parallel to the stress.

If the stress concentration due to the hatchway falls off with distance from the hatch as the reciprocal of the square root of distance, so that it has fallen to one-half of its maximum value at a distance of 12.5 mm, determine the average stress required in the deck plates for the crack to extend right across the deck.

Chapter 2: Answers

(Note: answers are rounded off to the appropriate numbers of significant figures.)

2.1   14.8 GPa

2.3   10.8 GPa, cleavage

2.5   0.32 nm

2.7   5.12 kJm^{-2}

2.9   115 T

2.11 1.12 mm

2.13 26.1 MPa, 14.0 MPa

Chapter 3: Problems

You are strongly recommended to solve these problems in the order given. Data needed will be found in tables in this and other chapters of the book.

3.1   The largest crack in the surface of a sample of tungsten wire is 2.7 μm deep. What is the strength of the piece of wire?

3.2   Compare the maximum loads that can be supported by a Kevlar fibre, a boron fibre, and a 0.10 mm diameter nylon fibre.

3.3   A boron fibre has a surface step which is 13 μm high. The radius at the inner corner of the step is 0.30 μm. Will the stress at the corner reach the theoretical strength (E/15) at a lower applied stress than the usual breaking strength of boron fibres?

3.4   Calculate the strength of a 1.3 μm diameter iron whisker which has a Frank–Read source extending across half the fibre diameter, and able to generate an infinite amount of slip on a plane with a normal at 5° to the fibre axis. For iron $b = 0.248$ nm and $E = 212$ GPa.

3.5   The compressive strength of a Kevlar fibre is about one-sixth of its tensile strength. What is the smallest diameter of rod on which the fibre can be wound without damage due to excessive compression?

3.6   Compare the minimum radii that can be used in the weaving of stiff carbon, boron, and E-glass fibres. If the glass fibres are spun, what is the minimum number of fibres that should be used in the yarn if it is to be used for weaving with the maximum possible flexure? Assume that the compressive and tensile strengths of these fibres are the same.

3.7   The distribution of surface cracks in a production run of glass fibres is such that each 10 cm length has, on average, 1 crack which is 1 μm long, 10 cracks which are 0.1 μm long, etc., so that the number of cracks of length a μm is l/a. Derive a relationship representing the strength of the fibres as a function of fibre length, and hence calculate the strength for a 3.0 mm length of glass.

3.8   An E-glass fibre 10.2 μm diameter and 3.0 cm long falls upon another fibre at a speed of 46 mms^{-1}. Assume that it is stopped by the fibre it hits, and as a result the stationary fibre is cracked. Calculate the surface area of crack produced if all the kinetic energy goes into producing the crack. If the crack has a constant depth around the whole of the fibre circumference, what would the strength of the fibre be reduced to as a result of the crack.

3.9   Calculate the maximum load that can be supported by a roving which consists of 1224 E-glass fibres and 15 boron fibres. (Hint: check the fibre strain; and assume that all the glass fibres have the same strength and all the boron fibres have the same strength.)

3.10 Some stiff carbon fibres, 7.8 μm in diameter, were added to S-glass fibres to increase the modulus of the yarn prior to weaving. The yarn contained 204 fibres each with 10.4 μm diameter. How many carbon fibres need to be added in order to increase the yarn strength marginally?

Chapter 3: Answers

3.1   1.03 GPa

3.3   Yes (2.0 GPa)

3.5   2.6 mm

3.7   2.0 GPa

3.9   48 kg

Chapter 4: Problems

4.1   Calculate the aspect ratio and the critical aspect ratio for strong graphite fibres 8.3 μm dia. and 2.0 mm long embedded in pure annealed aluminium. (Assume that the yield strength of the matrix is equal to its ultimate tensile strength.)

4.2   Calculate the maximum and average fibre stress for a glass fibre 10.2 μm dia. and 0.61 mm long embedded in a matrix that exerts a shear stress of 3.7 MPa at the fibre surface, when $m = 0.27$.

4.3   Calculate the epoxy matrix average stress and strain for a composite with aligned short steel wires, 0.34 mm diameter and 5.8 mm long, when the interfacial shear stress is 2.1 MPa, and $m = 0.68$.

4.4   What is the ratio of average fibre to matrix stress for a composite with aligned Kevlar fibres 12.1 μm diameter and 2.5 cm long with $m = 0.023$, in an epoxy matrix, with an interfacial shear stress of 5.8 MPa.

4.5   Calculate the strain for gross slip (ε_{1p}) for aligned boron-aluminium alloy with an aspect ratio of 11. The alloy yields at 270 MPa.

4.6   Calculate the modulus for continuous aligned S-glass-polycarbonate with $V_f = 0.64$.

4.7   Compare the strengths of epoxy resin and pure annealed aluminium reinforced with continuous aligned boron fibres, both with $V_f = 0.22$.

4.8   Estimate the composite stress at strains of 0.0023 and 0.0073 for an aligned chopped E-glass-polyester if the glass fibres have a diameter of 10.3 μm and a length of 0.34 mm, the matrix modulus is 2.63 GPa, and the interfacial shear stress is 11.7 MPa. $V_f = 0.57$.

4.9   Calculate the critical aspect ratio for Al_2O_3 whisker reinforced polyester if the coefficient of friction at the interface is 0.27 and the shrinkage stress is 21 MPa. Also estimate the stress required to give composites with these fibres a strain of 0.0031 when $V_f = 0.21$, and $E_m = 3.1$ GPa.

4.10   Calculate the strength of aligned Kevlar-epoxy made with fibres (a) 1.27 mm long and (b) 3.81 mm long. The fibre diameter is 11.8 μm, the interfacial shear stress is 6.2 MPa, and $V_f = 0.47$; for the resin, assume $\sigma_{my} = \sigma_{mu}$.

4.11   Calculate the stress and strain at the knee for aligned SiC whisker reinforced nickel. The whiskers are 101 μm long and have a cross-section of 0.93 pm^2. The nickel yields at a stress of 350 MPa, has a modulus of 200 GPa, and a volume fraction of 0.84. Assume that the whiskers can be treated as though they have circular cross-sections.

4.12   What is the ultimate tensile strength of aligned chopped stiff graphite-aluminium alloy, if the graphite is 0.53 mm long and 8.7 μm dia., and constitutes 57% of the total volume. The aluminium alloy yields at a stress of 110 MPa.

4.13   Calculate the strength of tungsten-nickel with $V_f = 0.63$ and $s = 2s_c$ and $s = s_c/2$. The nickel yields at a stress of 420 MPa.

4.14   Calculate E_1, E_2, G_{12} and v_{12} for aligned continuous E-glass-epoxy for $V_f = 0.71$, and hence determine the compliance matrix.

4.15   Determine the stiffness matrix for an aligned continuous lamina of boron-aluminium with $V_f = 0.67$.

4.16   Calculate the three strains $\varepsilon_1, \varepsilon_2$ and γ_{12} for a $V_f = 0.45$ aligned S-glass-polyester lamina stressed at right angles to the fibre direction, in the plane of the lamina. The modulus of the polyester is 2.47 GPa. The applied stress is 45 MPa.

4.17   Evaluate E_x, E_y, G_{xy}, and v_{xy} for aligned continuous tungsten-nickel for $\phi = 30°$, and $V_f = 0.62$. E_m for nickel is 200 GPa and $v_m = 0.31$.

4.18   Calculate the angle between stress and fibre direction for maximum strength for aligned continuous strong graphite-aluminium with $V_f = 0.69$ using the maximum stress theory. The aluminium has a yield strength of 94 MPa, and $\tau_{12u} = \tau_{my}$. What would the yield criterion theory predict for the strength at this angle, if $\sigma_{2u} = v_m\sigma_{mu}$?

4.19   Calculate the Young's and shear moduli for planar random Kevlar-epoxy with $s \gg s_c$ and $V_f = 0.38$.

4.20 Determine the stiffnesses $\bar{Q}_{11}$ and $\bar{Q}_{22}$ for a 3-layer E-glass-polyester laminate made from three pieces cut from the same sheet of aligned fibre material with $V_f = 0.76$, and oriented at angles of $0°$, and $\pm 60°$. The polyester has $E_m = 2.63$. Would this laminate twist when stressed?

Chapter 4: Answers

4.1 $s = 240$, $s_c = 47$

4.3   0.58 MPa, 0.00023

4.5   0.0071

4.7   0.79 GPa, 0.82 GPa

4.9   1320, 1.13 GPa

4.11 0.00175, 0.52 GPa

4.13 2.0 GPa, 0.77 GPa

4.15 $\begin{vmatrix} 316 & 42 & 0 \\ 42 & 166 & 0 \\ 0 & 0 & 61.6 \end{vmatrix}$ GPa

4.17 230 GPa, 230 GPa, 89 GPa, 0.29

4.19 22 GPa

Chapter 5: Problems

For all problems except 5.1 assume that the fibre packing is hexagonal.

5.1   Calculate P_f and n for an aligned S-glass-epoxy with the fibres packed in a regular array with fibres at the corners and centres of oblongs having sides in the ratio 2/1. $V_f = 0.63$.

5.2   Calculate the composite stress and strain at the slip point for continuous aligned Kevlar-epoxy with $V_f = 0.55$.

5.3   Calculate the maximum fibre stress at the slip point for annealed pure aluminium reinforced by aligned steel wires having an aspect ratio of 15. $V_f = 0.42$.

5.4   Calculate the composite stress at the matrix yield strain for aligned Al_2O_3 whisker reinforced chromium where the matrix yields at 620 MPa, $E_m = 279$ GPa, $v_m = 0.210$, and the whiskers have an aspect ratio of 51. $V_f = 0.32$.

5.5   The approximate equations can only be used when $m \leqslant \frac{1}{2}$. Calculate the maximum strain for which these equations can be used for (a) boron-aluminium alloy and (b) boron-epoxy. The aspect ratio of the fibres is 24, the aluminium yields at 84 MPa, and the epoxy adheres perfectly, has a residual compressive stress of 17 MPa, and a coefficient of friction at the interface of 0.32. $V_f = 0.50$, $v_1 = 0.34$.

5.6   Calculate the stress in an aligned SiC whisker reinforced polyester when the strain reaches the whisker breaking strain. Compare the result with the Rule of Mixtures strength of the composite. The adhesion coefficient is 0.32, the polyester has $E_m = 2.73$, $\sigma_{mv} = 77$ MPa, $V_m = 0.54$, and $\sigma_r = -14$ MPa. $\mu = 0.27$ and $v_1 = 0.34$. The whiskers are 6.1 mm long and have an effective diameter of 1.04 μm.

5.7   Determine the stress and strain at the slip point for aligned molybdenum-alumina. The molybdenum wires are 37 mm long and have a diameter of 0.43 mm, $V_f = 0.17$, $a = 0.023$, and $v_m = 0.20$.

5.8   Calculate the applied stress at the ultimate tensile strain of the cement, for aligned E-glass-cement with a shrinkage stress of 2.1 MPa, a coefficient of friction of 0.13, and $V_m = 0.73$. $\sigma_{mu} = 6.8$ MPa and $E_m = 21$ GPa. The adhesion coefficient is 0.27 and the fibre aspect ratio is 2.3 s_c. Would the result be significantly different if you assumed that the composite obeyed the Rule of Mixtures for modulus?

5.9   Calculate the composite stress for aligned strong graphite-glass with $V_f = 0.47$ when slip occurs over half the fibre length. The coefficient of friction is 0.11, $\sigma_{mu} = 83$ MPa, the adhesion coefficient is 0.24, the shrinkage stress is 12MPa, and the fibre diameter and lengths are 9.8 μm and 1.6 mm respectively. $v_1 = 0.22$ and $v_2 = 0.35$. The composite strain is 0.0126. (Hint: use equations 5.42 and 5.57.)

5.10 Show that the composite strain in question 5.9 is 0.0126. (This is greater than ε_{mu}. Such a strain can be generated in a composite without cracking, see Section 7.10.)

5.11 Determine the relative contributions of the elastic and slip stress transfer processes for continuous Kevlar-epoxy when the composite strain is equal to the fibre breaking strain. $a = 0.50$, and $V_f = 0.50$.

Chapter 5: Answers

5.1 3.9, 1.55

5.3 125 MPa

5.5 (a) 0.0030 (b) 0.0025

5.7 9.2 MPa, 24×10^{-6}

5.9 0.15 GPa

5.11 elastic/slip $= 1/14$

Chapter 6: Problems

6.1 Calculate the critical volume fractions for strong graphite–cement with $s \gg s_c$ when the fibres are (a) aligned, (b) random in a plane and (c) random in space. The cement alone fails at 6.8 MPa.

6.2 Use the same reasoning as for V_{fmin} to determine whether there is a minimum volume fraction for stiffening, or any other criterion for stiffening.

6.3 Calculate the critical volume fraction for planar random stiff graphite-epoxy, assuming that the resin is elastic up to the fibre breaking-point, and that the fibres are very long.

6.4 Use the stress–strain curve shown in fig. 1.1 to estimate V_{fmin} for aligned SiC whisker reinforced steel, where the whiskers have $s \gg s_c$.

6.5 Estimate V_{fmin} for the SiC–steel described in question 6.4, for whiskers with an aspect ratio of 84.

6.6 What is the smallest block length that could (theoretically) be observed when continuous aligned strong graphite-alumina is stressed to failure. $V_f = 0.43$, $\tau_i = 2.8$ MPa, and the fibre diameter is 8.7 μm.

6.7 Determine the stress, and the upper bound for the strain at which matrix cracking can occur for an aligned continuous glass-polyester which has been cooled to low temperature, so that $\sigma_{mu} = 12.3$ MPa and $E_m = 5.8$ GPa, and the resin is elastic to the breaking-point. Assume that the cooling does not affect the properties of the glass. $V_f = 0.31$.

6.8 Show that the upper bound for the composite failure strain is ε_{fu}, and calculate the difference between this and the lower bound, and calculate the failure stress for the composite described in question 6.7.

6.9 Calculate the stress and strain at the knee for continuous aligned boron-aluminium alloy with $\sigma_{my} = 130$ MPa and $V_f = 0.63$.

6.10 Determine the strain for a stress of 0.97 GPa applied to aligned continuous strong carbon–magnesium alloy with $V_m = 0.56$ and $\sigma_{my} = 0.34$ GPa.

6.11 Determine the strain for question 6.10 with the carbon fibres having an aspect ratio of 20.

6.12 Calculate the breaking strength and secant modulus at the breaking-point for an aligned strong carbon-aluminium alloy with $s = 158$, $\sigma_{my} = 140$ MPa, and $V_f = 0.38$.

6.13 Calculate the critical aspect ratio, the breaking strength and the secant modulus at the breaking-point for perfectly adhering aligned S-glass-epoxy with $v_1 = 0.34$, $\mu = 0.37$, $\sigma_r = -5.4$ MPa, $V_f = 0.58$ and $s = 2s_c$.

6.14 In question 6.13, determine the relative importance of the various terms by writing the expression for the strength in the form $\sigma_{1u} = E_L \varepsilon_{fu} - \dfrac{V_f \sigma_{fu} s_c}{2s} \left[\dfrac{1-A}{1+B} + C \right]$ and evaluating A, B and C. Is the matrix Poisson's shrinkage important?

6.15 Calculate the stress and strain at the slip point (ε_{1p}) for aligned tungsten–nickel where the tungsten has a diameter of 0.50 mm and a length of 5.0 mm; the nickel yields at 380 MPa and $V_f = 0.47$.

6.16 Estimate the critical aspect ratios, and the stress and strain at the slip point for perfectly adhering steel-polycarbonate with $s = 0.8 s_c$. The material has $\mu = 0.19$, $\sigma_r = 18$ MPa, and $V_f = 0.21$. Use v_m for v_1. If the material were adhering poorly, would the result be any different?

6.17 Determine the strength tensor for a poorly adhering continuous aligned regularly packed stiff carbon-polyethylene laminate with an interface that fails at a tensile stress of 1.4 MPa and fails in shear at 17 MPa. $V_f = 0.31$.

6.18 Compare the theoretical compressive strengths of continuous aligned E-glass- and Kevlar-polyester, given that the compressive strength of the Kevlar is 0.28 GPa, its compressive modulus is 70 GPa and the corresponding values for the polyester are 57 MPa and 3.8 GPa. $V_f = 0.55$. The compressive modulus of glass (and most other materials) is the same as the tensile modulus. For the glass use equation 6.41.

6.19 Show that the flaw distribution described in problem 3.7 can be used to estimate the maximum value of $\bar{\sigma}_f$, if the fibres break up into ever shorter pieces prior to composite failure. Hence estimate the strength of an aligned initially continuous glass-epoxy with $V_f = 0.64$ and fibre diameter of 10.2 μm. Assume that τ_i is a constant, equal to 7.3 MPa, and neglect elastic stress transfer.

Chapter 6: Answers

6.1   (a) 0.0024, (b) 0.0064, (c) 0.0121

6.3   0.052

6.5   0.0062

6.7   56 MPa, 0.0025

6.9   0.53 GPa, 0.00183

6.11 0.0095

6.13 1200, 2.84 GPa, 50 GPa

6.15 1.09 GPa, 0.0109

6.17 $\begin{vmatrix} 710 \\ 12 \\ 7.3 \end{vmatrix}$ MPa

6.19 1.00 GPa

Chapter 7: Problems

In this set of problems assume $G_m = G_f = 0$ and $\phi = 0$, unless values for these parameters are given in the problem, or are to be calculated from data given in the problem.

7.1   Calculate the work of fracture for aligned chopped S-glass- and E-glass-polyesters with a fibre length of 2.2 mm, diameter 9.8 μm, and $V_f = 0.54$. $\tau_i = 7.4$ MPa.

7.2   Estimate the work of fracture for aligned continuous boron-epoxy for fibre failure in the crack plane. $V_f = 0.74$, $d = 0.103$ mm, and $\tau_i = 4.6$ MPa.

7.3   Estimate the work of fracture for continuous aligned E-glass epoxy where the glass has flaws which permit fibre break-up into lengths such that $s = s_c$ near the crack plane. $V_f = 0.64$, $\tau_i = 7.9$ MPa, $d = 10.1$ μm.

7.4   Calculate the critical aspect ratio for a boron-aluminium alloy with $\sigma_{my} = 278$ MPa, and estimate the work of fracture for the aligned fibre composite with $V_f = 0.74$, $s = s_c$, $d = 0.103$ mm.

7.5   A composite is made with aligned S-glass rovings containing 2040 fibres. Each roving behaves as a single fibre bundle for fracture, with a coefficient of friction of 0.31. The shrinkage stress is 23 MPa, $V_f = 0.51$ and each fibre has a diameter of 10.4 μm. Estimate the work of fracture, assuming that in the bundle the fibres are hexagonally packed with the maximum volume fraction possible.

7.6   Calculate the work of fracture for aligned SiC whisker-nickel, where the length of the whiskers is 0.68 mm and their effective diameter is 1.25 μm. The nickel has a yield stress of 570 MPa and $V_m = 0.62$.

7.7   What is the crack opening displacement required to generate the work of fracture for the SiC-nickel described in question 7.6?

7.8   Calculate the crack opening displacement required for generation of the work of fracture for the boron-aluminium described in question 7.2, and show that it is one-quarter of the value for question 7.4, although the works of fracture are the same.

7.9   What is the critical aspect ratio for E-glass fibres crossing a crack obliquely, with $\phi = 60°$, in an epoxy matrix with $\tau_i = 4.6$ MPa.

7.10 Calculate the work of fracture for continuous aligned boron-aluminium alloy with $\phi = 20°$. The yield stress of the alloy is 0.41 GPa, the fibre diameter is 0.105 mm and $V_f = 0.63$.

7.11 Calculate the work of fracture for aligned Al_2O_3 whisker-nickel. The effective whisker diameter is 1.07 μm and the length is 0.18 mm, $V_f = 0.37$, $\sigma_{mu} = 480$ MPa, and $\phi = 25°$.

7.12 Show that, for brittle fibres at moderate angles ($\phi \leqslant 45°$), the $\tan \phi$ term normally has a much larger effect on the work of fracture than the $0.72 \, \varepsilon_{fu} \tan^2 \phi$ term. To do this determine the ratio of the amount each changes $G_{fb\phi}$, for the fibre or whisker having the largest ε_{fu} and smallest A, in the softest matrix normally used for high performance composites (epoxy resin). (Remember that $(l-x)^3 \simeq 1 - 3x$ for $x \ll 1$).

7.13 Estimate the work of fracture for aligned stiff carbon-epoxy. The fibres have a diameter of 7.9 μm and a length of 2.0 mm. $V_f = 0.54$, $\tau_i = 4.7$ MPa, $\phi = 35°$.

7.14 What fraction of the total work of fracture is done by plastic bending of aligned steel wires when $\phi = 60°$, if $\sigma_{fu} = \sigma_{fy} = 1.43$ GPa. The matrix is epoxy resin with $\tau_i = 23$ MPa, and the fibres have a length of 17 mm and a diameter of 0.55 mm. (Do not include G_f or G_m).

7.15 Compare the fibre works of fracture for random molybdenum-alumina with fibre lengths of (a) 70 mm and (b) 34 mm. The fibre diameter is 0.37 mm, $\tau_i = 6.3$ MPa, $V_f = 0.17$ and $\phi = 55°$. Consider only the work arising from flexure during pull-out.

7.16 What percentage difference does including the work of fibre fracture make to the solutions of problem 7.15? The work of fracture for the molybdenum fibres is 43 kJm^{-2}.

7.17 In question 7.15b the fibres debond completely at each crack face. If the debonding work is equal to the work of fracture of the matrix, by what percentage does this increase the work of fracture?

7.18 If U_f for steel is 143 MJm^{-3}, what would be the work of fracture for continuous aligned steel 0.56 mm diameter in an epoxy resin with $\tau_i = 23$ MPa and $V_m = 0.36$.

7.19 Make a rough estimate of the volume fraction of E-glass required to double ε_{mu} in a cement with τ_i 2.7 MPa, $\sigma_{mu} = 7.4$ MPa, and $G_m = 0.38$ Jm^{-2}. The diameter of the glass is 9.8 μm (Hint: assume $V_m \simeq 1$).

7.20 Calculate the work of fracture of pure annealed aluminium with a deformed zone at each fracture surface which is 5.4 mm thick, and a strain of 1.07, and compare this with the matrix work, $V_m G_m^*$, when reinforced by stiff carbon fibres 7.9 μm diameter with $V_f = 0.68$. Compare also the work involved when the fibres fracture in the crack plane. (Hint, for calculation of G_m see question 2.11.)

Chapter 7: Answers

7.1   0.33 MJm^{-2}, 0.33 MJm^{-2}

7.3   0.39 MJm^{-2}

7.5   7.3 MJm^{-2}

7.7   0.34 mm

7.9   340

7.11 2.8 kJm^{-2}

7.13 0.17 kJm^{-2}

7.15 (a) 1.5 MJm^{-2} (b) 0.97 MJm^{-2}

7.17 0.021 %

7.19 0.19

Chapter 8: Problems

Assume the platelets are square unless otherwise indicated.

8.1 What size should a SiC platelet be in order to be fully effective when symmetrically packed in a polyester resin with $\tau_i = 4.7$ MPa. The SiC is 1.17 μm thick.

8.2 Calculate the size required for a mica flake with imperfect edges to give a strength efficiency factor of 0.60 when non-symmetrically packed in nylon with $\tau_i = 3.8$ MPa. The mica is 2.73 μm thick.

8.3 Calculate the strength of an aluminium diboride-aluminium alloy, with $s = 30$ when (a) symmetrically packed, and (b) unsymmetrically packed. The aluminium alloy yields at 228 MPa, and the aluminium diboride has a strength of 5.8 GPa and $V_p = 0.67$.

8.4 Calculate the stress and secant modulus at the breaking-point of unsymmetrically packed SiC-nickel with $V_p = 0.73$, $\sigma_{my} = 470$ MPa, $E_m = 200$ GPa and $s = 65$.

8.5 What is the slope of the stress–strain curve at the breaking-point of (a) symmetrically packed and (b) unsymmetrically packed mica-epoxy with imperfect edges with $s = 150$, $\tau_i = 5.8$ MPa, $V_p = 0.73$.

8.6 A bar of symmetrically packed aluminium boride-epoxy 1.35 m long is deformed, and increases in length by 7.8 mm. Calculate the stress and strain of the composite, and the secant modulus. $\sigma_{pu} = 6.2$ GPa, $E_p = 530$ GPa and $V_p = 0.83$.

8.7 Calculate the strain in a non-symmetrically packed mica-polyethylene with $s = 300$, $V_p = 0.86$ and $\tau_i = 4.3$ MPa when the applied stress is 0.31 GPa.

8.8 Estimate the work of fracture of silicon carbide platelet reinforced epoxy with $\tau_i = 7.8$ MPa, $V_p = 0.68$, $t = 1.32$ μm, and $2L = 0.53$ mm. $G_m \simeq 0$.

8.9   Glass platelets have been tried (unsuccessfully) as reinforcements. Their strength is 0.13 GPa (i.e. less than twice that of sheet-glass) and $E_p = 70$ GPa. Estimate the toughness of a glass-epoxy made with round glass discs of this strength, having a diameter of 12.7 mm, a thickness of 51 μm, and with $V_p = 0.64$. $\tau_i = 6.7$ MPa. $G_m \simeq 0$. (Hint: a good approximation is obtained if it is assumed that the platelets are square, but have the same surface area.)

8.10 A silicon carbide platelet-epoxy is required with $s > s_c/2$, and a work of fracture of 400 kJm^{-2}. What should $2L$ be for this, for the composite whose other parameters are as given in question 8.8?

Chapter 8: Answers

8.1   2.5 mm

8.3   (a) 1.22 GPa (b) 1.22 GPa

8.5   166 GPa, 12 GPa

8.7   0.0017

8.9   0.65 kJm^{-2}

Chapter 9: Problems

9.1   Assuming that the unidirectional fibre composite in Table 9.2 does not obey the Rule of Mixtures for strength because of damage to the fibres during manufacture, calculate the size of flaws in the glass that must be present. (Neglect the matrix contribution, and assume an E-glass-epoxy, with a strength fraction calculated on the basis of data in Table 3.2.)

9.2   In the case of the random mat in Table 9.2, it is likely that the fibres retain only about 60% of their strength due to damage in processing. Also they cannot transfer stress efficiently due to their short lengths. Making these assumptions, estimate the aspect ratio of the fibres. Express your result as a fraction of the critical aspect ratio. (Hint: superpose the effect due to randomness, Section 4.3.4, on that due to short length Section 4.1. Neglect the matrix contribution.)

9.3   Calculate the compressive strengths of E-glass-epoxy, and boron-epoxy assuming that failure occurs when the matrix yields. Hence determine the ratios of the theoretical compressive strengths to Rule of Mixtures tensile strengths at $V_f = 0.60$. Assume $\sigma_{my} = \sigma_{mu}$. Do these results agree with those obtained in practice?

9.4   Calculate the ratio of effective compressive strength to effective tensile strength for Kevlar when used to reinforce epoxy, using Rule of Mixtures expressions, and the data in Section 9.2.1.

9.5   Assume the Rule of Mixtures applies to the carbon-epoxy in Table 9.3, and hence calculate V_f from S_{11}. Then determine how close the elastic constants S_{22}, S_{12} and S_{66} come to the theoretical values, expressing your answers as the ratios of theoretical to experimental values. Assume stiff carbon was used.

9.6   Determine the fibre strength from the data in Table 9.3, assuming $V_f = 0.54$. Do the off-axis strengths and shear strengths agree with the theory?

9.7   What would the effective strengths of boron, carbon, and S-glass have to be for the works of fracture in Table 9.4 to be determined by stress redistribution (G_{fb})? Assume that $\tau_i = 6.0$ MPa in all cases, and that the high modulus carbon was used.

9.8   If the data in Table 9.3 is for the same material as that for the carbon-epoxy in Table 9.4 what would be the value for the fracture toughness (K_{1c}) of the material?

9.9   Use the data in Table 9.4 to design a Kevlar-stiff carbon hybrid epoxy with $E_1 = 100$ GPa and $G = 100$ kJm^{-2}. Determine the volume fractions of each required, neglecting any matrix contribution.

9.10 Design a strong graphite-E glass hybride epoxy with a strength of 1.4 GPa and a modulus of 140 GPa. Determine the volume fractions, neglecting the matrix contribution.

9.11 The mica-epoxy described in Table 9.7 had an aspect ratio of about 100. Calculate τ_i from the composite flexural strength, assuming $s < s_c$. Hence estimate the lower bound for σ_{pu}.

9.12 Use Fig. 9.32 to estimate the minimum length of lap joint needed to transmit 10% of the breaking load for a continuous aligned E-glass epoxy laminate which obeys the Rule of Mixtures. Consider two cases (a) 4 layers and (b) 9 layers, where each layer has a thickness of 1.2 mm and all the fibres are parallel. $V_f = 0.47$. Would you expect such a lap joint to be able to transmit 10% of the breaking load for the nine layer laminate?

Chapter 9: Answers

9.1   24.7 nm

9.3   Glass, 0.51, boron, 2.9; no

9.5   0.54, 0.90, 1.06, 1.13

9.7   1.65 GPa, 4.8 GPa, 4.0 GPa

9.9   Kevlar 0.39, carbon 0.13

9.11 $\tau_i = 5.9$, $t = 12$ nm

Chapter 10: Problems

In all the following problems neglect elastic stress transfer effects.

10.1   If the low result for the longitudinal strength of the carbon-nickel alloy in Table 10.1 was due to fibre break-up during manufacture, what would be the fibre length for $\sigma_{my} = 440$ MPa, $d = 8.3$ μm and $V_f = 0.6$.

10.2   In question 6.19 the fibres broke up into lengths such that the average fibre stress was

$$\bar{\sigma}_f = \frac{3}{4}\left(\frac{\tau_i \mathbf{K}^2}{d}\right)^{1/3}$$

where

$$\mathbf{K}^2 = BE_f G_f /\pi(1 - v^2)$$

and B = constant. Assume that the high result for carbon-aluminium in Table 10.1 is due to the same effect, and hence calculate the work of fracture for the fibres; considering the strong form of graphite. The fibres were well bonded in pure annealed aluminium with $d = 8.3$ μm and $V_f = 0.55$. Because they were more flawed than the glass of question 6.19, $B = 20 \times 10^3$ (instead of 10^5).

10.3   The strength of the W–Ni alloy shown in Table 10.2 is 1.05 GPa, the fibres have a diameter of 0.47 mm, and the nickel yields at 0.62 GPa. Use these data, and data in the Table to estimate the strain in the nickel at the fracture surfaces, and the ratio of the depth of deformed nickel with fibres present to that in the absence of fibres. Significant contributions to the work of fracture come from the matrix fracture work, and G_{fb}.

10.4   What fraction of the work of fracture of the boron-aluminium alloy shown in Table 10.2 comes from stress redistribution? Assume that the boron is weakened in the manufacture of the composite, to an extent which can be calculated from the data in Table 10.1. The fibre diameter is 0.102 mm and $\sigma_{mu} = 320$ MPa.

In the following two questions use the approximate equations for stresses given in Table 10.4 together with equations 10.11 and 10.12.

10.5   What change of temperature is required to cause a 10% reduction in the strength of initially stress-free continuous aligned tungsten-nickel alloy composite with $V_f = 0.50$ and a matrix yield strength which remains constant at 620 MPa in the temperature range in question. $\sigma_{fu} = 1.41$ GPa and $E_m = 200$ GPa.

10.6   What is the lowest temperature at which one of the matrix stresses (σ_{mr}, $\sigma_{m\theta}$, σ_{mz}) would exceed the yield stress for silica-aluminium. The composite is initially stress-free at 20 C, has $V_f = 0.50$, and the yield stress of the matrix falls linearly with temperature, from 310 MPa at 20 C to 110 MPa at 300 C.

Chapter 10: Answers

10.1   0.066 mm

10.3   0.102, 0.082

10.5   280 C increase

Chapter 11: Problems

11.1   Choose fibres from Table 3.3 which are suitable for reinforcing (a) pyrex glass, (b) glass ceramic (c) alumina (d) silicon and (e) silicon nitride without surface roughening being necessary, assuming that the fibre thermal expansion is isotropic.

11.2   Calculate the minimum depth of roughness needed for silicon carbide fibres, 87 μm diameter, to be used to reinforce silicon. Will the strength of the fibres be reduced, and if so, by how much, if the roughening process also makes cracks which are five times deeper than the troughs needed for stress transfer.

11.3   If account is taken of the different coefficients of expansion in the radial and axial directions, would smooth carbon fibres be suitable for reinforcing pyrex glass?

11.4 The carbon-pyrex in Fig. 11.4 obeys the Rule of Mixtures, but the carbon appears to be rather weak. Assuming the stiff graphite was used, what notch length must be present in the carbon to account for the observed fibre strength if the work of fracture for the graphite is $6.0\ Jm^{-2}$?

11.5 Calculate the theoretical work of fracture, assuming that interface shear can take place, for the SiC-Si described in sections 11.1.4 and 11.1.5, for $V_f = 0.7$. Assume that the work of fracture for SiC-Si is described in Sections 11.1.4 and 11.1.5 is entirely due to stress redistribution, G_{fb}. Hence calculate τ_i. The fibre diameter is $14\ \mu m$.

11.6 Calculate the effective strengths of the glass, steel, asbestos, and Kevlar fibres from the flexural strengths listed in Table 11.4, assuming $s \gg s_c$. In which cases do synergistic effects appear to be taking place.

11.7 What would s/s_c have to be for the glass-cement in Table 11.4 to account for the flexural strength, assuming that the effect of the fibre random orientation can be superposed on the aspect ratio effect. If the glass behaves as a bundle of 204 fibres, and has a length of 5.1 cm, what is the value of τ_i.

11.8 Estimate the interfacial shear stress for Kevlar needed to give the toughness results in Table 11.4, for an aspect ratio of $2s_c$. The diameter of the fibres is $12.3\ \mu m$.

11.9 Calculate the interfacial shear stress for the glass-cement giving the toughness result in Table 11.4. For the cement use $\tau_{my} = \tau_{mu} = \sigma_{mu}$ (flexure). For the glass consider two cases: (a) the fibres acting as bundles of 204 fibres and (b) the fibres acting singly. The fibre diameter is $10.1\ \mu m$ and the length is 5.1 cm (Note that s_{cr} cannot be greater than s_c; the approximations give slight errors when $\phi_c \simeq \pi/2$).

11.10 Estimate the change in τ_i due to 20 years immersion in water for glass-cements (a) based on flexural strength (see problem 11.7) when the fibres act as a bundle and (b) based on toughness when the fibres act singly (see problem 11.9). Are the results mutually compatible? (Note use the percentage reductions given in Table 11.4, but apply them to the data in Table 11.3.)

Chapter 11: Answers

11.1 (a) possibly tungsten, (b) none, (c) S-glass, (d) possibly tungsten, (e) none.

11.3 No; radial separation would occur

11.5 11 MPa

11.7 1.6, 7.9 MPa

11.9 (a) 62 MPa (b) 4.1 MPa

Chapter 12: Problems

12.1 A business man flies regularly on a 250 km non-stop journey at a height of 900 m over the sea. The plane consumes energy at a rate of $95\ J s^{-1} kg^{-1}$ to keep it aloft and moving at its cruising speed of $210\ km\,h^{-1}$. By calculating the fuel consumption determine which material would be better for the wings and fuselage if the choice is aluminium or fibreglass. Fibreglass has a density of $1.75\ gm\,cm^{-3}$, but more of it is needed to preserve the flexural rigidity of the structure, since its Youngs modulus is only 35 GPa. When made of aluminium the volume of material needed is $0.051\ m^3$. The other parts weigh 230 kg, and the business man weighs 67 kg. The heat of combustion of the kerosene fuel is $52.8\ GJ\,m^{-3}$; its weight may be neglected; the engine is 27% efficient. Calculate the amount of fuel used in each case.

12.2 Develop a criterion for a material to be used for floors in passenger carrying jet aircraft. The aircraft can make journeys totalling 10 million km before the floor needs replacing. The floor needs flexural rigidity, and any weight saved can be used to carry extra passengers giving an increase in income of $0.05 per passenger km. The average weight of a passenger plus baggage and seat is 90 kg. Compare an aluminium alloy costing $2 kg^{-1} with graphite-epoxy costing $50 kg^{-1}, and having $V_f = 0.75$. The density of the epoxy is $0.98\ Mgm^{-3}$.

12.3 Show that the elastic energy stored in a bar in tension is $\frac{1}{2}$ stress x strain x volume. Hence, determine whether a better material than rubber can be used to drive model aircraft. For simplicity, compare each in tension, and neglect the weight of the system needed to convert linear movement to rotation. Consider silica-, boron-, and carbon-epoxy instead of natural rubber with a density of $1.13\ M\,gm^{-3}$; V_f cannot exceed 0.75 without loss of properties.

12.4 Compressed gas is being considered as an energy store. Show that the energy stored is PdV for pressure P and change in volume dV, and hence that, with a perfect gas (for which PV is constant) the energy stored in compressing it from pressure P_1 to P_2 is $P_2 V_2 \ln(P_2/P_1)$ where V_2 is the volume of the container. Suppose that a gas turbine can be designed to work at 90% efficiency over a pressure range from full pressure P_2 to $P_2/2$. (At lower pressures its efficiency becomes negligible.) It will work at a remote site, so the gas storage vessel must be as light as possible. The vessel will be a long buried tube, so that the stress in the tube wall is $P_2D/2t$, and uniaxial, where D = tube diameter and t = wall thickness. Derive a figure of merit for the material, and compare the performance of the composites considered in the previous question with the aluminium alloy the Table 1.1.

12.5 A rotating long thin tube is under stress due to centrifugal forces. For a peripheral rotation velocity of v, and a mass M per unit tube external surface area, the forces exert an equivalent pressure equal to $2Mv^2/D$ where D is the tube diameter. Calculate the stress in the tube, and show that the kinetic energy stored in the tube (half the tube mass, multiplied by the square of its velocity) cannot exceed the product of its ultimate tensile strength and the volume of the material in the tube. Now flywheels have been considered for energy storage in moving vehicles. Here the volume of the ring material has to be kept to a minimum. Compare the materials in the two previous questions on the basis of maximum energy stored per unit volume of material.

12.6 Compare the efficiency of the various forms of energy storage in questions 12.3, 12.4 and 12.5 with that of kerosene, on a volume basis. (This is important for road transport.) In your answer list the various methods in descending order of effectiveness, giving the energy stored in Jm^{-2} for the most effective in each category.

12.7 A cylindrical pressure vessel with hemispherical ends, unlike a long pipe, has to withstand axial stresses as well as circumferential (hoop) stresses. Show, by suitably dividing such a vessel into two halves, in two different ways, that the hoop stress is exactly twice the axial stress. Hence, using the maximum stress theory in Section 4.3.3, calculate the fibre winding angle, $\pm\phi$, needed for efficiently winding the cylindrical part of the vessel. Consider the case $\sigma_{2u} < 0.7\ \tau_{12} < 0.5\sigma_{1u}$. Express your answer as the angle between the fibres and the hoop stress.

12.8 The Ford Motor Company are making a car with a carbon-polymer body, in order to reduce petrol consumption. The carbon-polymer has a density of 1.45 Mgm^{-3}. The body uses 0.035 m^3 of material. The other parts of the car weigh 420 kg. How much petrol is used and what percentage saving can be made, compared with a steel body, using the same volume of material, in an average daily journey to work. In this average journey there is a driver weighing 72 kg, no luggage, and no passangers. The distance travelled is about 10 km, and the journey takes 25 minutes, of which 30% is spent accelerating, 30% decelerating, 20% waiting at traffic lights, and 20% travelling at constant speed. Assume that the acceleration is always at a rate of 1.6 ms^{-2} up to a speed of 50 $km\,h^{-1}$. The engine is 20% efficient when accelerating, but requires an energy input of 20 kJs^{-1} to keep it turning over when idling and 95 kJs^{-1} for travelling at 50 $km\,h^{-1}$. The heat of combustion of petrol is 48.5 $GJ\,m^{-3}$.

12.9 A reasonably good cross country skier can work at a rate of 230 watts for long periods. On the level he travels 3.1 m with each leg movement. If we assume that all the work goes into accelerating the ski and the leg below the knee from rest to twice and skiers average velocity, calculate and skiers speed on the level with skies made (a) from wood with $\sigma_{1u} = 34$ MPa and $\rho = 0.86$ and (b) strong graphite-epoxy with $V_f = 0.75$. The skier weighs 70 kg, and each foot and leg below the knee weighs 2.7 kg. The ski is 2.30 m long, and for simplicity we will assume constant, oblong section, with a width (b_1) of 5.4 cm and a thickness determined by the material strength. Thus it should just support the skier when only the two ends are in contact with the ground.

12.10 A support for use for X-ray therapy is fixed at one end and must deflect as little as possible when a patient is on it. It must also absorb X-rays as little as possible. The bed is made as a sandwich, which can be as thick as needed to keep flexure to a minimum. The filling resists shear but does not contribute to the bending moment. Its X-ray absorbtion can be neglected. The surface skins take all the tensile and compressive stresses. (Consequently the stresses in them can be calculated by equating the external moments with the internal moment coming entirely from the skins). Use a Rule of Mixutres expression for the calculation of X-ray absorbtion, using the atomic numbers as a measure of the absorbtion per unit thickness. The stresses in the skins cannot exceed half the ultimate tensile strength. Derive a criterion of excellence, and determine the relative positions of silica-, boron- and carbon-epoxy with $V_f = 0.75$, and aluminium. The atomic numbers are: boron 5, carbon 6, aluminium 13. For silica use 10 (the average for Si and two O's) and for epoxy use 3.5 (an average for hydrogen, oxygen and carbon).

Chapter 12: Answers

12.1 Fibreglass better; 3.21: aluminium; 3.41.

12.3 In order of excellence: silica-epoxy, 9.4 Jg^{-1}; rubber, 2.5 Jg^{-1}; graphite-epoxy, 0.79 Jg^{-1}; boron-epoxy 0.50 Jg^{-1}.

12.5   In order of excellence: silica-epoxy, 2.2; boron-epoxy, 1.3; carbon-epoxy, 1.06; aluminium alloy, 0.33; rubber, 0.016.

12.7 $35°$

12.9   (a) 3.8 ms^{-1}, (b) 4.6

Index